LARGE SETS OF TRIPLE SYSTEMS AND RELATED DESIGNS

Applied Discrete Mathematics and Theoretical Computer Science

LARGE SETS OF TRIPLE SYSTEMS AND RELATED DESIGNS

Kang Qingde

SCIENCE PRESS
NEW YORK, LTD.

Responsible Editor: Lü Hong

Published by Science Press New York, Ltd.
84-04 58th Avenue, Elmhurst, NY 11373, U.S.A.

Printed in USA

ISBN 1-880132-13-3 (New York)
ISBN 7-03-004490-8/ O • 773 (Beijing)

SERIES EDITORS' PREFACE

Combinatorics is one of the oldest branches of mathematics. Through its many diverse applications, combinatorics has developed from the point of studying finite structures to that of considering an extremely broad range of discrete structures which often do not arise in traditional pure mathematics

During the last twenty–five or more years, there has been an extremely fruitful interface between discrete mathematics and the new subject of theoretical computer science. This has developed primarily through the study of theoretical computer science, and applications of the approach taken in discrete mathematics in its applications to such areas as optimization and operations research.

In this series we present monographs and texts that will make the exciting new ideas in this area accessible to a broad audience.

Ronald V. BOOK
Ding-Zhu DU

Preface

Combinatorics is the fastest-growing area of modern mathematics. It contains the following parts: combinatorial enumeration, combinatorial design, graph theory, combinatorial algorithm, combinatorial optimizing, and so on.

Starting with two famous recreational probloms, Euler's problem of the Thirty six officers and Kirkman's schoolgirl problem, and stimulated by the needs of computer science, digital communication, modern business management, the subject of design theory has grown into a large branch of combinatiorial mathematics.

Large set of designs is an important topic of design theory, which has some valuable applications to experimental design, coding theory and so on. Though the topic has acquired numerous great advances in recent years, due to its difficulty, the progress is not more than expected.

The purpose of this book is to present the recent developments in the field of the large sets of block designs, especially by Chinese mathematicians. The main topics are about latest advances on the large sets of triple systems. As well as, some researches about large sets of other designs are also introduced.

In Chapter 1, some essential concepts and the background of Algebra, combinatorial designs, triple systems, graphs and large sets will be introduced. What we are going to talk about in Chapter 2 is the studying process about large sets of Steiner triple systems, to which Lu Jiaxi and other scholars made enormous contribution. In Chapters 3–5, three classes of large sets of oriented triple systems are dealt with, i.e., Mendelsohn triple systems, directed triple systems and hybrid triple systems, among which the main work has been done by us. In Chapter 6, some preliminary results about the large sets of Mendelsohn designs are presented. In Chapter 7, we will discuss a few related designs, including Golf design (i.e., the large set of idempotent symmetric Latin squares), large set of GDD, and the solution about two open problems. Lastly, a biographical sketch of Lu Jiaxi is followed as the appendix of this book.

In this field, many mathematicians such as R. H. F. Denniston, A.

Rosa, C. C. Lindner, Luc Teirlinck and Lu Jiaxi have made their remarkable contributions. Besides, C. J. Colbourn, M. J. Colbourn, Y. M. Chee, T. Etzion, E. S. Kramer, D. R. Stinson, A. P. Street, M. J. Sharry, S. Schreiber, S. A. Vanstone, R. M. Wilson and Wu Lisheng have also done a lot about the research. We, the author and his students, have just followed them up with further study in this book.

This book is expected to be of some help to those who are interested in the subject, and we hope our initial job will benefit their research work and the development of the field.

Acknowlegements:

I grant my thanks to Prof. Du Dingzu for his inspiration and kind suggestion to write the book. I would also express my thanks to Prof. Luc Teirlinck for his manuscripts [Teir 6,7,8].

I wish to express my gratitude to the Hebei Normal College for providing the better condition and support to write this book.

I am very grateful to several young friends: Gao Tianzeng (for his help in translating the appendix), Bi Quanqi, Lei Jianguo, Chang Yanxun, Zhao Dongxing, Zhang Jie and Tian Zihong (for their help in typing and proofreading).

Contents

Chapter 1

Preliminaries

1.1 Some background about Algebra

This section contains some definitions that most readers will be acquainted with already—we add it for completeness only.

Let S and T be two sets. We denote by $S \times T$ the set of all ordered pairs (x, y) with $x \in S$ and $y \in T$. This set is called *Cartesian product* of S and T. Similarly, we can define Cartesian products of $S_1, S_2, \cdots, S_{k-1}$ and S_k, $S_1 \times S_2 \times \cdots \times S_k$. When $S_1 = S_2 = \cdots = S_k = S$ the products $S \times S \times \cdots \times S$ are called *power set* of S. Any subset of $S \times T$ is called a *correspondence* between S and T. A *mapping* from S to T is a correspondence between S and T such that to each $s \in S$ there corresponds exactly one $t \in T$. If f is a mapping, usually one writes

$$f : S \longrightarrow T \quad \text{and} \quad s \longmapsto t.$$

An important special case of mapping is that of one-to-one mapping, also called a *bijection*. Here there corresponds to each $t \in T$ just one $s \in S$. Given a mapping $f : S \longrightarrow T$ we shall denote the mapping from the power set of S to the power set of T defined by

$$A \longmapsto \{f(a) : \ a \in A\}$$

by the same symbol f.

By a *monoid* we understand a set S with an element e and a mapping $f: S\times S \longrightarrow S$ such that if $f(x,y)$ is the result of applying f to the pair $(x,y)\in S\times S$ then

$$\begin{cases} f(x,f(y,z)) = f(f(x,y),z), & \forall x,y,z\in S,\\ f(e,x) = x = f(x,e), & \forall x\in S.\end{cases}$$

The mapping f is called a (*binary*) *operation* on S and denoted by xy (instead of $f(x,y)$). The element e is called its *unity element*. An element x of a monoid S is said to be *invertible* if there exists $x'\in S$ such that $xx' = x'x = e$. If x is invertible, the corresponding x' is unique and is denoted by x^{-1}. By a *group* one understands a monoid in which every element is invertible. If the operation of a group is commutative (i.e., $xy = yx$, $\forall x,y$) then the group is called *abelian*. A *subgroup* of the group G is a subset of G, which is a group relative to the operation in G. Given a group G and a subgroup H, for any $x\in G$ the subset of $G: xH = \{xh:\ h\in H\}$ (resp. $Hx = \{hx:\ h\in H\}$) is called a *left* (resp. *right*) *coset* of H in G. Clearly,

$$xH = yH \text{ (or } Hx = Hy) \text{ if and only if } x^{-1}y \text{ (or } xy^{-1}) \in H.$$

There is a partition of the group G in the form

$$G = \bigcup_x xH \quad (\text{or } G = \bigcup_x Hx).$$

For a finite group G containing n elements, and any $x\in G$, there exists a positive integer m with $m \mid n$ such that $x^m = e$ and $x^k \neq e$ ($\forall 0<k<m$), where e is the unity element of G. Call the integer m the *order* of the element x. Let S be an arbitrary subset of a group G. Then there exists a unique smallest subgroup of G containing S. We shall denote this subgroup by (S) and call it the *subgroup generated by* S. In the case $S = \{x_1,\cdots,x_t\}$ we also write $(x_1,\cdots,x_t)$ for this subgroup. A *cyclic group* is a group generated by a single element.

Let X be a finite set containing n elements $1,2,\cdots,n$. A *permutation* ξ of X is a bijection from X to X, and is usually denoted by

$$\xi = \begin{pmatrix} 1 & 2 & \cdots & n\\ i_1 & i_2 & \cdots & i_n\end{pmatrix},$$

where $i_k = \xi(k), 1 \leq k \leq n$. All such permutations of X form a group, which is called the *symmetric group* on X and is denoted by $Sym(X)$, brief S_n. It is well known that S_n (or $Sym(X)$, if X is a v−set) contains just $n!$ elements. The permutation which maps i_j to i_{j+1} ($0 \leq j \leq k-1$, the subscript mod k) and fixes each element in $X \backslash \{i_0, i_1, \cdots, i_{k-1}\}$ is called a *cyclic permutation* of length k and is denoted by $(i_0, i_1, \cdots, i_{k-1})$. It is well known that each permutation can be written as a product of disjoint cyclic permutations.

A *ring* (with identity) is a set R with two binary opertions, $(x, y) \mapsto x + y$, called addition, and $(x, y) \mapsto xy$, called multipliation, such that (i) R is an abelian group under the addition; (ii) R is a monoid under the multiplication; (iii) The addition and the multiplication are related by the distributive laws:

$$(x+y)z = xz + yz \quad \text{and} \quad z(x+y) = zx + zy, \quad \forall x, y, z \in R.$$

The unity element of the additive group is called the *zero* element of the ring and is usually denoted by 0.

For a given positive integer n, all integers are separated into n classes: $\bar{0}, \bar{1}, \cdots, \overline{n-1}$, where

$$\bar{k} = \{in + k; \quad i \in Z\}, \quad k = 0, 1, \cdots, n-1,$$

and Z is the set containing all integers. Provided with the two operations defined by

$$\overline{k_1} + \overline{k_2} = \overline{k_1 + k_2} \quad and \quad \overline{k_1} \cdot \overline{k_2} = \overline{k_1 k_2},$$

the set $\{\bar{0}, \bar{1}, \cdots, \overline{n-1}\}$ forms a ring, which is called the *residue class ring modulo n* and is denoted by Z_n.

A ring is called a *field*, if its non-zero elements (for its multiplication) form a commutative group. The unity element of the multiplicative group is usually denoted by 1. It is well known that if a finite field contains v elements, then v must be a prime power, and conversely, if v is a prime power then up to isomorphism there is a unique field with v elements. A finite field is usually denoted $GF(p^n)$ or F_{p^n} where p is a prime and n is a positive integer. The multiplicative group of a finite field F is a cyclic group , a generator g of which is called a *primitive element* of

$F^* = F\backslash\{0\}$. So the set of non-zero elements of F_{p^n} can be written as $\{1, g, g^2, \cdots, g^{p^n-2}\}$.

A so-called *quasigroup* is a set Q with a binary operation, multiplication $\circ$, such that both maps

$$l_Q : x \longmapsto a \circ x \quad \text{and} \quad r_Q : x \longmapsto x \circ a$$

are permutations of Q for each $a \in Q$. It is denoted by $(Q, \circ)$. A quasigroup $(Q, \circ)$ is called *idempotent* if $a \circ a = a$ for each $a \in Q$.

In this book, we shall use special notations about set: $I_n = \{0, 1, ..., n-1\}$, a set with no operation; G, a group (with an operation); $Z_n = \{0, 1, ..., n-1\}$, a ring (with two operations); and F_q, a field. And, for a real number α, the symbol $\lceil \alpha \rceil$ represents the lowest integer not less than α, and the symbol $\lfloor \alpha \rfloor$ represents the greatest integer not greater than α.

1.2 Combinatorial Designs

In combinatorics the subject of design theory, roughly speaking, is the study of incidence structures satisfying certain numerical requirements. One may define a *combinatorial design* to be a way of selecting subjects from a finite set in such a way that some specified conditions are satisfied. It contains many sub-branches such as block designs, Latin squares, difference sets, Hadamard matrices.

In general, a *block design* is a pair $(X, \mathcal{B})$, where X is a set and $\mathcal{B}$ is a collection of subsets of X, such that some conditions are satisfied. The elements of X will be called *points* (or *elements*), those of $\mathcal{B}$ *lines* (or *blocks*). A sub-collection of $\mathcal{B}$ is called a *parallel class* if it forms a partition of X. A design $(X, \mathcal{B})$ is called *resolvable* when $\mathcal{B}$ can be partitioned into parallel classes.

The classical type of block designs is the so-called $BIBD$ (*balanced incomplete block design)*. A block design $(X, \mathcal{B})$ is called a (v, k, λ)-$BIBD$ (or $B[k, \lambda; v]$), if X is a v-set, each block of $\mathcal{B}$ is a k-subset of X and each 2-subset of X appears exactly in λ block of $\mathcal{B}$. Here, the parameters v, k and λ are named *order*, *block size* and *index* of the $BIBD$. It is easy to see that $\mathcal{B}$ contains $b = \frac{\lambda v(v-1)}{k(k-1)}$ blocks and each element of X appears exactly in $r = \frac{\lambda(v-1)}{k-1}$ blocks of $\mathcal{B}$. Thus, (v,k,λ)-$BIBD$ is named (v,b,r,k,λ) -$BIBD$ also. And, a (v,k,λ)-$BIBD$ exists only if

$$\begin{cases} \lambda v(v-1) \equiv 0 & (\text{mod } k(k-1)) \\ \lambda(v-1) \equiv 0 & (\text{mod } (k-1)) \end{cases}$$

A resolvable $B[k,\lambda;v]$ is usually denoted by $RB[k,\lambda;v]$.

A generalization of $BIBD$ is so-called *t-design*. A t-(v,k,λ) design is a pair $(X,\mathcal{B})$, where X is a v-set and $\mathcal{B}$ is a collection of some k-subsets of X such that each t-subset of X appears exactly in λ blocks of $\mathcal{B}$. By the definition, a (v,k,λ)-$BIBD$ is a 2-(v,k,λ) design. In a t-(v,k,λ) design $(X,\mathcal{B})$, $\mathcal{B}$ contains $b = \lambda\binom{v}{t}/\binom{k}{t}$ blocks and every s-subset of X belongs to exactly $r_s = \lambda\binom{v-s}{t-s}/\binom{k-s}{t-s}$ blocks, when $s \le t$. Thus a t-(v,k,λ) design is an s-(v,k,λ_s) design for $s \le t$; where $\lambda_s = \binom{v-s}{t-s}/\binom{k-s}{t-s}$. Sometimes, a $t-(v,k,\lambda)$ design is denoted by $S_\lambda(t,k,v)$.

Another generalization of the $BIBD$ is the PBD (*pairwise balanced design*). A block design $(X,\mathcal{B})$ is called a (v,K,λ)-PBD (or $B[K,\lambda;v]$), if X is a $v-$set and the size of each block of $\mathcal{B}$ belongs to the set K such that every 2-subset of X is contained in exactly λ blocks of $\mathcal{B}$. When $K = \{k\}$ (i.e. $|K| = 1$), the (v,K,λ)-PBD will be a (v,k,λ)-$BIBD$. We shall denote by $B(K)$ the set of all integers v for which there exists a PBD $B[K,1;v]$. A set K is said to be *PBD–closed* if $B(K) = K$. Pairwise balanced designs are of fundamental importance in combinatorial theory and have been used extensively in the construction of other types of combinatorial designs.

A *group-divisible design* $GD[K,\lambda,M;v]$, briefly GDD, is a triple $(X, \mathcal{G}, \mathcal{A})$, where X is a $v-$set, $\mathcal{G}$ is a partition of X into m_i-subsets (called *groups*, $m_i \in M$) and $\mathcal{A}$ is a collection of k_i-subsets of X (called *blocks*, $k_i \in K$) such that $|B \cap G| \le 1$ for all $B \in \mathcal{A}$ and $G \in \mathcal{G}$ and such that any 2-subset P of X with $|P \bigcap G| \le 1$ for all $G \in \mathcal{G}$, is contained in exactly λ blocks of $\mathcal{A}$. The *type* of a $GD[K,\lambda,M;v]$ is $m_1^{r_1} m_2^{r_2} \cdots m_t^{r_t}$, if $M = \{m_1, m_2, \cdots, m_t\}$ and there are r_i groups with size m_i in $\mathcal{G}$, $1 \le i \le t$. A $(v,K,\lambda) - PBD$ is a $GD[K,\lambda,\{1\};v]$, and a $GD[K,1,M;v] =$

$(X, \mathcal{G}, \mathcal{A})$ is a $(v, K \cup M, 1) - PBD$. When $K = \{k\}$, $M = \{m\}$ and $v = km$, the $GD[k, \lambda, m; km]$ is named *transversal design* and is denoted by $TD[k, \lambda; m]$. A $TD[k, 1; m]$ can be written as $TD(k, m)$ briefly. In a $TD(k, m) = (X, \mathcal{G}, \mathcal{A})$, $|X| = km$, $\mathcal{G} = \{G_1, G_2, \cdots, G_k\}$, $X = \bigcup_{i=1}^{m} G_i$, $|G_i| = m, 1 \leq i \leq k$, and $|B| = k$, $|B \cap G_i| = 1$ for any $B \in \mathcal{A}$ and $G_i \in \mathcal{G}$. There is an importance relationship between transversal design and another design – Latin square, which will be stated below.

A *Latin square* of order n is defined to be an n-by-n array made out of n distinct symbols with the property that each of the n symbols occurs exactly once in each row and in each column of the array. Let $A = (a_{ij})_1^n$ and $B = (b_{ij})_1^n$ be two Latin squares of order n. If the n^2 ordered pairs (a_{ij}, b_{ij}) for $1 \leq i, j \leq n$ are distinct, then A and B are called *orthogonal*, and B is called an *orthogonal mate* of A. In a Latin square $A = (a_{ij})_1^n$ of order n, if there are n positions (i_k, j_k) in distinct rows and distinct columns such that the n^2 entries $a_{i_k j_k}$ for $1 \leq k \leq n$ are distinct, then these positions are called a *transversal* of A. Latin square A of order n possesses an orthogonal mate if and only if A has n disjoint transversals. A Latin square $A = (a_{ij})_1^n$, $1 \leq a_{ij} \leq n$, is called *idempotent* if $a_{ii} = i$ for any i, and is called *symmetric* if $a_{ij} = a_{ji}$ for any i and j. It is easy to see that an (idempotent) Latin square of order v is equivalent to an (idempotent) quasigroup containing v elements. It is well known that there exists a pair of orthogonal Latin squares of order v if and only if $v \neq 2$ and 6, and there exists an idempotent Latin square of order v if $v \geq 3$.

An *orthogonal array* $OA_\lambda(t, k, v)$ is an array with k columns on a v-set X such that if we run t fingers down any t columns, we find every ordered t-tuple of (not necessarily distinct) elements of X exactly λ times. If $\lambda = 1$, then the subscript λ is not specified. Especially $OA(2, k, v)$ is denoted by $OA(v, k)$ briefly. The existence of an $OA(v, k)$ is equivalent to the existence of a set of $k - 2$ mutually orthogonal Latin squares of order v, and is equivalent to the existence of a $TD(k, v)$.

1.3 Triple systems

The simplest nontrival $BIBD$'s are the $B[3,1,v]$, i.e. $(v,3,1)$-$BIBD$, which is called *Steiner triple system*, briefly $STS(v)$. A resolvable $STS(v)$ is called *Kirkman triple system*, briefly $KTS(v)$. In general, a block design with block size $k = 3$ is named a *triple system* and denoted by $TS(v,\lambda)$ with index λ. By definition of $BIBD$, in a $TS(v,\lambda)$, every block $\{x,y,z\}$ is an unordered triple, which may be regarded as a collection of three unordered pairs $\{x,y\},\{y,z\}$ and $\{z,x\}$. If the collection is taken to consist of ordered rather than unordered pairs, then we will get the extensions of unordered triple systems — oriented triple systems.

There are two types of oriented triple. Let x,y and z be distinct elements of a given set X. A *cyclic triple* is a collection of three ordered pairs $(x,y),(y,z)$ and (z,x), which is denoted $\langle x,y,z\rangle$ (or $\langle y,z,x\rangle$ or $\langle z,x,y\rangle$). A *transitive triple* is a collection of three ordered pairs $(x,y),(y,z)$ and (x,z), which is denoted by (x,y,z). In graphical viewpoint, an unordered triple $\{x,y,z\}$ can be viewed as a triangle consisting of three edges $\{x,y\},\{y,z\}$ and $\{z,x\}$, but both cyclic triple $\langle x,y,z\rangle$ and transitive triple (x,y,z) are directed triangles by assigning edge's direction on the underlying triangle $\{x,y,z\}$ in two manners.

A *Mendelsohn* (resp. *directed*) *triple system* with index λ is a pair $(X,\mathcal{B})$, where X is a v-set and $\mathcal{B}$ is a collection of some cyclic (resp. transitive) triples from X such that each ordered pair of distinct elements of X belongs to exactly λ triples of $\mathcal{B}$. This system is denoted by $MTS(v,\lambda)$ (resp. $DTS(v,\lambda)$). N. S. Mendelsohn first studied the system MTS under the name "cyclic triple system" which unfortunately can be confused with triple systems having a cyclic automorphism. And the system DTS was first introduced by S. H. Y. Hung and N. S. Mendelsohn. In some articles the system DTS was once named as TTS (*transitive triple system)*. But , in this book, we rather use the notation DTS to avoid confusion with twofold triple system (i.e. $TS(v,2)$).

A *hybrid triple system* with index λ is a pair $(X,\mathcal{B})$, where X is a v-set and $\mathcal{B}$ is a collection of both cycltriple system with index λ is a pair $(X,\mathcal{B})$, where X is a v-set and $\mathcal{B}$ is a collection of both cyclic and transititve triples from X such that every ordered pair of distinct elements of X appears in precisely λ triples of $\mathcal{B}$. This system is denoted by $HTS(v,\lambda)$. C. J. Colbourn, W. R. Pulleyblank and A. Rosa first introduced this generalization of MTS and DTS.

Two $MTS(v,\lambda)$s (or $DTS(v,\lambda)$s, or $HTS(v,\lambda)$s, or $STS(v)$s) are said to be *isomorphic* if there exists a bijection of the respetive point sets that preserves blocks. An *automophism* of a triple system is an isomorphism with itself. There are the similar concepts for other designs also.

1.4 Graphs and decomposition into cycles

A *graph* G is a pair (V,E), where $V = V(G)$ is a finite nonempty set of elements called *vertices* and $E = E(G)$ is a finite set of unordered pairs of distinct elements of $V(G)$ called *edges*. The number $|V(G)|$ is called the *order* of G. Often, the edge joining the vertices u and v will be denoted as uv or $\{u,v\}$. If $e = uv$ is an edge of G, then u and v are called the two *ends* of e. The ends of an edge are said to be *incident* with the edge, and vice versa. Two vertices (resp. edges) that are incident with a common edge (resp. vertex) are said to be *adjacent*. A set of vertices of which every pair is adjacent is called a *clique*. A *k-clique* is a clique with k vertices.

A *cycle* of length t in a graph G is a sequence of t distinct vertices $v_1v_2\cdots v_t$ such that v_iv_{i+1} (for $1 \le i \le t-1$) and v_tv_1 are edges of G. If the length of a cycle is t, then it is called *t-cycle*. In a graph of order v, a cycle of length v is called a *Hamiltonian cycle*.

There are two graphs that are of particular importance in design theorey. The graph K_v is the *complete graph* on v vertices and is defined by the fact that every pair of vertices are adjacent. A *multipartite graph* is one in which the vertex set can be partitioned into sets (called *parts*) in such a way that each edge joins a vertex in two different parts. The *complete multipartite graph* $K_{n_1,\cdots,n_t}$ is an multipartite graph where the vertex set is partitioned into t parts having sizes $n_1,\cdots,n_t$, where every vertex is adjacent to every vertex in a different part.

A *directed graph* (or *digraph*) is a graph in which each edge has been assigned a direction. If the direction of an edge $\{x,y\}$ is assigned as $x \rightarrow y$

then the edge (called *arc* for digraph) is denoted by (x, y). In a digraph, for a given vertex, there are two kinds of arcs : *in-arcs* and *out-arcs*. A *directed cycle* (or *circuit*) of length t in a digraph D is a sequence of t distinct vertices $v_1 v_2 \cdots v_t$ such that (v_i, v_{i+1}), for $1 \leq i \leq t-1$, and (v_t, v_1) are arcs of D. If the length of a circuit is t, then we call it a *t-circuit* . A *Hamiltonian circuit* in D is a circuit containing all the vertices of D. If each vertex of D has just one in-arc and one out-arc, then D is the union of some of arc disjoint circuits. A *complete symmetric digraph* K_v^* is a graph with v vertices in which for any two distinct vertices x, y there are arcs (x, y) and (y, x).

For an $STS(v) = (X, \mathcal{B})$, Steiner triple system of order v, if each triple of $\mathcal{B}$ is considered as a triangle (i.e. 3-cycle), then the system $\mathcal{B}$ can be regarded as a partition of the complete graph K_v with vertices set X into edge disjoint 3−cycles. Thus, in graph theory, the system $STS(v)$ is called *3-cycle system* of order v. In general, an *m-cycle system* of order v is a pair (S, C), where C is an edge disjoint collection of m-cycles which partition the K_v based on the v-set S.

Similarly, for digraph, we have so-called *m-circuit system* of order v, which is a pair (S, C), where C is an arc disjoint collection of m-circuits which partition the complete symmetric digraph K_v^* based on the v-set S. So, by the definition, the $MTS(v)$, Mendelsohn triple system of order v, is a 3-circuit system of order v.

More generally, given a graph (resp. digraph) G, a *G-design* $(X, \mathcal{B})$ of order v is a v-set X together with a family $\mathcal{B}$ of graphs (resp. digraphs), each isomorphic to given G, such that for some constant index λ each edge (resp. arc) of the complete graph K_v (resp. complete symmetric digraph K_v^*) on X is covered by precisely λ members of $\mathcal{B}$. By the definition, the system $STS(v)$ is a G-design of order v with index 1, where G is a triangle, and the system $MTS(v, \lambda)$ is also a G-design of order v with index λ , where G is a cyclic triangle. But, the system $DTS(v, \lambda)$ is a G-design, where G is a transitive trangle. By the way, as a generalization of G-design, the graph G can be substituted with a collection $\mathcal{G}$ of some graphs and each member of $\mathcal{B}$ will be isomorphic to some member of $\mathcal{G}$. Then, the system $HTS(v, \lambda)$ will be a $\mathcal{G}$-design, where $\mathcal{G}=$ { cyclic triangle, transitive triangle}.

1.5 Large sets of designs

Maybe the first to investigate large sets of designs was Cayley (1850), who showed that a large set of Steiner triple systems of order 7 does not exist and who together with Sylvester, following Kirkman's schoolgirl problem, posed such question: Is there a large set of Kirkman triple system of order 15 ?

In general, a *large set* of G-designs is a collection $\{(X, \mathcal{B}_i)\}_i$ of G-designs on the same point set X, $|X| = v$, such that every subgraph of K_v (or K_v^*) on X that is isomorphic to the given graph G is a member of precisely one family $\mathcal{B}_i$. If G is taken as k-clique, then the large set of G-designs of order v with index λ is just the large set of $B[k, \lambda; v]$, i.e. $\{(X, \mathcal{B}_i)\}_i$, where each $(X, \mathcal{B}_i)$ is a $B[k, \lambda; v]$, and the collection $\{\mathcal{B}_i\}_i$ is a partition of all k-blocks of X. This large set is denoted by $LS_\lambda(2, k, v)$. If G is taken as cyclic triangle (or transitive), we will get the large set of $MTS(v, \lambda)$ (or $DTS(v, \lambda)$), which is denoted by $LMTS(v, \lambda)$ (or $LDTS(v, \lambda)$). Similarly, we can refer to $LHTS(v, \lambda)$, $LKTS(v)$ and $LGDD \cdots$.

Let's examine the matter from other view-point. We denote by $D_\lambda(2, k, v)$ the maximum number of mutually disjoint $B[k, \lambda; v]$. Since there are $\frac{\lambda v(v-1)}{k(k-1)}$ k-blocks in a $B[k, \lambda; v]$ and the total number of k-blocks from the point set is $\binom{v}{k}$, therefore $D_\lambda(2, k, v) \leq \binom{v-2}{k-2}/\lambda$. If the equality occurs, then there exists an $LS_\lambda(2, k, v)$, or else $LS_\lambda\ (2, k, v)$ does not exist. Of course, there is the case $D_\lambda(2, k, v) < \binom{v-2}{k-2}/\lambda$, for example $D(2, 3, 7) = 2 < \binom{7-2}{3-2} = 5$, thus $LSTS(7)$ does not exist. Two large sets $\{(X, \mathcal{A}_i)\}_i$ and $\{(Y, \mathcal{B}_j)\}_j$ of the same type and the same order are said isomorphic if there exists a bijection from X onto Y such that each $\mathcal{A}_i$ is mapped onto some $\mathcal{B}_j$.

Recently, Luc Teirlinck has developed the theory about large set and raised so-called *Large sets with holes*. This development has played and will play important influence on the studing for large sets. In Chapter 3 and 5, we will talk of it.

Chapter 2

Large sets of Steiner triple systems

2.1 History

A Steiner triple system of order v (briefly $STS(v)$) on a v-set X is a $B[3,1;v] = (X, \mathcal{B})$, where $|\mathcal{B}| = v(v-1)/6$. It is well known that there exists an $STS(v)$ if and only if $v \equiv 1$ or $3 \pmod 6$ and $v \geq 3$ (refer to [Kir 1]). A *large set of disjoint* $STS(v)$, briefly $LSTS(v)$, is a partition of all 3-subsets (triples) of X into $STS(v)$. We denote by $d(v)$ the maximum number of pairwise disjoint $STS(v)$'s, and denote by $d^*(v)$ the maximum number of pairwise disjoint and isomorphic $STS(v)$'s. It is easy to see that $d^*(v) \leq d(v) \leq v-2$ for $v \geq 3$, and there exists an $LSTS(v)$ if and only if $d(v) = v - 2$.

In 1850, A. Cayley showed that $d(7) = 2$ (see [Cay]), i.e. there exists no $LSTS(7)$. This is the first work about $LSTS$, maybe is also the beginning investigation about large sets of designs. In the same year, T. P. Kirkman showed that $d(9) = 7$, i.e. there exists an $LSTS(9)$, which is the first nontrivial $LSTS$ ($d(3) = 1$ is trivial). In 1972, J. Doyen proved that [Do]

$$d^*(6m+1) \geq \begin{cases} \frac{m}{2} & (\text{ if } m \text{ even}) \\ m & (\text{ if } m \text{ odd}) \end{cases},$$

$$d^*(6m+3) \geq \begin{cases} 4m+1 & (\text{ if } m \equiv 0, 2 \bmod 3) \\ 4m-1 & (\text{ if } m \equiv 1 \bmod 3) \end{cases} \quad \text{and}$$

$$d(2v+1) \geq d(v) + 2 \text{ for } v \geq 7.$$

In 1975, Kotzig proved $d(6m+1) \geq 2m-1$ for odd m. In 1973, Luc Teirlinck proved that $d(3v) \geq d(v)+2v$ $(v \geq 3)$. By his recursive method, the first infinite family— $LSTS(3^n)$ was obtained. As well as, in 1975, A. Rosa presented another recursive theorem: $LSTS(v) \rightarrow LSTS(2v+1)$ for $v \geq 7$. Besides, S. Schreiber (1973), R. M. Wilson (1974) and R. H. F. Denniston (1974) had successively gotten the existence of $LSTS(v)$ for the following orders:

v =51, 73, 75, 81, 91, 105, 129, 153, 163, 169, 193, 201([Sch], [Wil 2])

v =13, 15, 19, 21, 25, 31, 33, 43, 49, 61, 69 ([Den 2]).

Until 1982, the existence problem of $LSTS(v)$ had been still open, and was far from complete solution. In 1983–1984, Lu Jiaxi, a Chinese amateur mathematical researcher, published his six papers in *Journal of Combinatorial Theory*, see [Lu 1—6]. After making tremendous efforts, he proved the following theorem.

Theorem 2.1.1. If $v \equiv 1$ or 3 (mod 6), $v > 7$ and $v \notin \{$ 141, 283, 501, 789, 1501, 2365 $\}$, then $d(v) = v - 2$.

Thereupon, the existence problem of $LSTS$ had been basically solved except six possible orders. In 1989, Luc Teilinck pointed the fact that the set $\{n :$ there exists an $LD^*(n)\}$ has the PBD-closed property (see [Teir 6]), where $LD^*(n)$ is a type of auxiliary design introduced by Lu, which will be discussed in the next section. By the property, Teirlinck has solved the existence of $LSTS(v)$ for the remaining six orders, therefore the existence spectrum for $LSTS$ has been finally completed.

In this Chapter, we will mainly state Lu's contribution. To shorten his proof in [Lu 1—6] and to present an useful tool for other designs, the $LD-$design and LD^*-design will be discussed in detail in Section 2.2, where the existence conclusion of LD and LD^* have been given by us.

In Section 2.3, we introduce some auxiliary designs AD, AD^*, AD^{**} and their large sets LAD, LAD_1, LAD_2 and LAD_3, which appear in [Lu 2] and [Lu 4], and relate their (including LD's) important role to construct $LSTS$. In Section 2.4, a proof of the existence theorem of $LSTS$ is stated, which has been already shortened by us even though this shortening is not yet enough. As well as, the existence result about $LS_\lambda(2,3,v)$ and another way to construct $LSTS$ are introduced.

2.2 LD and LD^* designs

2.2.1 Preface and concepts

To construct $LSTS$, Lu Jiaxi introduced a kind of auxiliary design—LD design in [Lu 3], which played an important role in his work. And, by his imagination [Lu 7], the design shall be able to solve the $LSTS$ problem for the remaining six orders. Besides, in our opinion, it is also worth to investigate the LD design as a new kind of combinatorial design. Actually, in the next chapter (Section 3.4) we will see that the LD-design can be used to solve the existence problem of other designs.

An *LD-design*, $LD(n) = LD[X]$, is a quadruple $(X, \mathcal{L}^1, \mathcal{L}^2, (\mathcal{L}_x)_{x\in X})$ satisfying the following conditions:

(C1) X is a n-set.

(C2) For each $j \in \{1,2\}, \mathcal{L}^j$ is an $OA(2,4,n)$ on X.

(C3) For each $x \in X$, $\mathcal{L}_x$ is an $OA(2,3,n-1)$ on $X\backslash\{x\}$.

(C4) There is an element $c_0 \in X$ such that $(x,x,x,c_0) \in \mathcal{L}^1 \cap \mathcal{L}^2$ for all $x \in X$.

(C5) Any $(x_0,x_1,x_2) \in X^3$ is either contained in an $\mathcal{L}_x$ $(x \in X)$ or in an $\overline{\mathcal{L}}^j (j \in \{1,2\})$, where $\overline{\mathcal{L}}^j = \{(x_0,x_1,x_2);$ there is an $x_3 \in X$ with $(x_0,x_1,x_2,x_3) \in \mathcal{L}^j\}$, $j=1,2$.

By definition, it is a kind of design that consists of several transversal designs with a certain balanced connection. By the equivalency between the transversal design and the orthogonal Latin squares, the definition can be stated as:

$LD(n) = LD[X] = \{ A^1, A^2, A_x : x \in X\}$, where $X = \{0, 1, 2, \ldots, n-1\}$ and

i) Each A_x is an $n \times n$ array $(a^x_{st})_0^{n-1}$ with $a^x_{sx} = a^x_{xt} = \infty$ $(o \le s, t \le n-1)$ and $\tilde{A}_x = (a^x_{st})_{s \ne x, t \ne x}$ is a Latin square over $X \backslash \{x\}$.

ii) Each $A^j = (b^j_{st})_0^{n-1}, j = 1, 2$, is an idempotent Latin square over X having n disjoint transversals.

iii) For any $s, t \in X, \{a^x_{st}\}_{x \in X} \bigcup \{b^j_{st}\}_{j \in \{1,2\}} = X \bigcup \{\infty\}$, where the element ∞ appears twice (if $s \ne t$) or element x appears twice (if $s = t = x$).

It is easy to verify the equivalence of both definitions. Here we only point out that

$$A_x = (a^x_{st}) \text{ if and only if } \mathcal{L}_x = \{(s, t, a^x_{st})\},$$

$$A^j = (b^j_{st}) \text{ if and only if } \mathcal{L}^j = \{(s, t, b^j_{st}, *)\}.$$

As an example, we give $LD(4) = LD[I_4]$:

$$A_0 = \begin{bmatrix} \infty & \infty & \infty & \infty \\ \infty & 3 & 2 & 1 \\ \infty & 2 & 1 & 3 \\ \infty & 1 & 3 & 2 \end{bmatrix}, \quad A_1 = \begin{bmatrix} 2 & \infty & 0 & 3 \\ \infty & \infty & \infty & \infty \\ 0 & \infty & 3 & 2 \\ 3 & \infty & 2 & 0 \end{bmatrix},$$

$$A_2 = \begin{bmatrix} 3 & 1 & \infty & 0 \\ 1 & 0 & \infty & 3 \\ \infty & \infty & \infty & \infty \\ 0 & 3 & \infty & 1 \end{bmatrix}, \quad A_3 = \begin{bmatrix} 1 & 0 & 2 & \infty \\ 0 & 2 & 1 & \infty \\ 2 & 1 & 0 & \infty \\ \infty & \infty & \infty & \infty \end{bmatrix},$$

$$A^1 = \begin{bmatrix} 0 & 3 & 1 & 2 \\ 2 & 1 & 3 & 0 \\ 3 & 0 & 2 & 1 \\ 1 & 2 & 0 & 3 \end{bmatrix}, \quad A^2 = \begin{bmatrix} 0 & 2 & 3 & 1 \\ 3 & 1 & 0 & 2 \\ 1 & 3 & 2 & 0 \\ 2 & 0 & 1 & 3 \end{bmatrix}.$$

A class of special LD-designs, LD^*-designs, was also introduced by Lu. The LD^*-design satisfies the additional condition besides (C1)—(C5) above:

(C6) The set $\mathcal{L}^0 = \{(x_0, x_1, x_2, x_3) : (x_0, x_2, x_3) \in \overline{\mathcal{L}}^1$ and $(x_1, x_2, x_3) \in \bar{\mathcal{L}}^2\}$ is an $OA(2, 4, n)$ on X.

Obviously, the existence of $LD^*(v)$ implies the existence of $LD(v)$. It is not difficult to see that the above-mentioned example is also an $LD^*(4)$.

In [Lu 3] and [Lu 5], Lu gave a sequence of directed and recursive construction for $LD(n)$ and $LD^*(n)$. But the existence spectrum of LD and LD^* has not completed yet. The following are all results about LD and LD^* given by Lu.

(L1) For any positive integer $\alpha > 1$, there exists an $LD^*(2^\alpha)$.

(L2) For $n \equiv 5 (\mathrm{mod}\ 8)$, there exists an $LD(n)$. For $n = 5$ even an $LD^*(n)$ exists.

(L3) For $n \equiv 7 (\mathrm{mod}\ 12)$, there exists an $LD(n)$. For $n = 7, 19$ even an $LD^*(n)$ exists.

(L4) For $n \equiv 11$ (mod 12) except n=23, 47, 59, 83, 107, 167, 179, 227, 263, 299, 347, 383, 719, 767, 923, and 1439, there exists an $LD(n)$. For $n = 11$ even an $LD^*(n)$ exists.

(L5) If there exist both an $LD^*(n_1)$ and an $LD(n_2)$, then there exists an $LD(n_1 n_2)$.

(L6) For odd prime p and positive integer α, if there exists an $LD(p^\alpha)$ then there exists an $LD(3p^\alpha)$.

(L7) For prime power $q > 4$ and positive integer n, if there exists an $LD(1+n)$ then there exists an $LD(1+qn)$.

(L8) For odd prime power $q \geq 7$ and $0 \leq t \leq 2^{r-1} - 3 (r > 2)$, if there exist both an $LD(1+q)$ and an $LD(1+(q-1)t)$ then there exists an $LD(1+(q-1)t+2^r q)$.

(L9) For odd prime power $q, q' \geq 7$ and $0 \leq t \leq (2^{r-1}-3)\ (q'-3), r > 2$, if there exist both an $LD(1+qq')$ and an $LD(1+(q-1)t)$ then there exists an $LD(1+(q-1)t+2^r qq')$.

(L10) For $n = 2, 3, 6$ there does not exist an $LD(n)$.

Luc Teirlinck, in his paper to solve the remaining six orders of $LSTS$ [Teir 6], has pointed that the set of orders n for which an $LD^*(n)$ exists has the PBD-closed property. By this inspiration, we have basically completed the existence spectrum for LD and LD^*. And, this work of ours will allow to substantially shorten Lu's proof although certain parts of the proof remain complicated. Our conclusion is

Theorem 2.2.1. For integer $v > 1$, if $v \in F$ then $LD(v)$ does not exist, if $v \notin M \bigcup F$ then there exists an $LD(v)$ and if $v \notin N \bigcup F$ then there exists an $LD^*(v)$, where $F = \{2, 3, 6\}$, $M = \{10, 14, 18, 23, 26, 30, 38, 42, 62, 74, 90\}$ and $N = M \bigcup \{12, 15, 27, 66\}$.
To prove the theorem we need some Lemmas, which are shown in Subsections 2.2.2 and 2.2.3.

2.2.2 Directed constructions

Lemma 2.2.2. There exists an $LD^*(v)$ for $v = 5, 7, 9, 11$ and 19; and there is no $LD(v)$ for $v = 2, 3$ and 6.

Construction.

(1) Let $F = GF(5)$. Firstly, given an ordered triple group T as follows:
$T = \{(0,0,2), (0,1,1), (0,3,0), (0,2,3), (1,0,3), (1,1,0), (1,3,1), (1,2,2)\}$.
Define

$\mathcal{L}_0 = \{(2^{2t+i}, 2^{2t+j}, 2^{2t+k}) : \ (i, j, k) \in T, t = 0, 1\}$;
$\mathcal{L}_x = \mathcal{L}_0 + x$ for any $x \in F \backslash \{0\}$;
$\mathcal{L}^1 = \{(x + y, y, 4x + y, x) : \ x, y \in F\}$;
$\mathcal{L}^2 = \{(3x + y, y, 4x + y, x) : \ x, y \in F\}$.

Then $(F, \mathcal{L}^1, \mathcal{L}^2, (\mathcal{L}_x)_{x \in F})$ is an $LD^*(5)$.

(2) Let $F = GF(7)$. Given a cyclic triple group
$C = \{\langle 4,4,0 \rangle, \langle 1,1,2 \rangle, \langle 0,1,3 \rangle, \langle 0,3,1 \rangle\}$.
Define

$\mathcal{L}_0 = \{(3^{2t+i}, 3^{2t+j}, 3^{2t+k}) : \ \langle i, j, k \rangle \in C, t = 0, 1, 2\}$;
$\mathcal{L}_x = \mathcal{L}_0 + x$ for any $x \in F \backslash \{0\}$;
$\mathcal{L}^1 = \{(4x + y, y, 6x + y, x) : \ x, y \in F\}$;
$\mathcal{L}^2 = \{(2x + y, y, 6x + y, x) : \ x, y \in F\}$.

Then $(F, \mathcal{L}^1, \mathcal{L}^2, (\mathcal{L}_x)_{x \in F})$ is an $LD^*(7)$.

(3) Let g be a primitive element of $F = GF(3^2)$, which satisfies the equation $g^2 = g + 1$. Take an ordered triple group T as follows:
$T = \{$(0,4,0), (1,0,1), (0,7,7), (1,6,6), (1,4,3), (0,6,1), (1,7,4), (0,1,2), (0,5,3), (0,2,4), (1,2,0), (1,5,7), (1,3,2), (0,3,6), (0,0,5), (1,1,5)$\}$.
Define

$\mathcal{L}_0 = \{(g^{2t+i}, g^{2t+j}, g^{2t+k}) : \quad (i,j,k) \in T, t = 0,1,2,3\};$

$\mathcal{L}_x = \mathcal{L}_0 + x$ for any $x \in F\backslash\{0\};$

$\mathcal{L}^1 = \{(x+y, g^2x+y, g^7x+y, y) : \quad x,y \in F\};$

$\mathcal{L}^2 = \{(gx+y, x+y, g^2x+y, y) : \quad x,y \in F\}.$

Then $(F, \mathcal{L}^1, \mathcal{L}^2, (\mathcal{L}_x)_{x\in F})$ is an $LD^*(9)$.

(4) Let $F = GF(11)$. Given an ordered triple group

T = {(0, 0, 7), (0, 1, 8), (0, 3, 2), (0, 5, 6), (0, 7, 0), (0, 9, 4), (1, 1, 9), (1, 3, 7), (1, 6, 0), (1, 7, 3), (1, 9, 1), (2, 0, 3), (3, 0, 0), (4, 0, 9), (5, 4, 0), (6, 0, 5), (7, 1, 1), (7, 8, 0), (8, 0, 1), (9, 2, 0) }.

Define

$\mathcal{L}_0 = \{(2^{2t+i}, 2^{2t+j}, 2^{2t+k}) : \quad (i,j,k) \in T, t = 0,1,2,3,4\};$

$\mathcal{L}_x = \mathcal{L}_0 + x$ for any $x \in F\backslash\{0\};$

$\mathcal{L}^1 = \{(7x+y, y, 3x+y, x) : \quad x,y \in F\};$

$\mathcal{L}^2 = \{(10x+y, y, 3x+y, x) : \quad x,y \in F\}.$

Then $(F, \mathcal{L}^1, \mathcal{L}^2, (\mathcal{L}_x)_{x\in F})$ is an $LD^*(11)$.

(5) Let $F = GF(19)$. Given a cyclic triple group

$C = \{\langle 0,0,9\rangle, \langle 0,1,2\rangle, \langle 0,3,6\rangle, \langle 0,5,10\rangle, \langle 0,7,14\rangle, \langle 0,11,4\rangle, \langle 0,13,8\rangle, \langle 0,15,12\rangle, \langle 0,17,16\rangle, \langle 1,3,7\rangle, \langle 1,7,3\rangle, \langle 9,9,1\rangle\}.$

Define

$\mathcal{L}_0 = \{(2^{2t+i}, 2^{2t+j}, 2^{2t+k}) : \quad \langle i,j,k\rangle \in C, 0 \le t \le 8\};$

$\mathcal{L}_x = \mathcal{L}_0 + x$ for any $x \in F\backslash\{0\};$

$\mathcal{L}^1 = \{(11x+y, y, 18x+y, x) : \quad x,y \in F\};$

$\mathcal{L}^2 = \{(7x+y, y, 18x+y, x) : \quad x,y \in F\}.$

Then $(F, \mathcal{L}^1, \mathcal{L}^2, (\mathcal{L}_x)_{x\in F})$ is an $LD^*(19)$.

(6) Since there exists no $OA(2,4,2)$ and $OA(2,4,6)$ hence both $LD(2)$ and $LD(6)$ do not exist. And, it is easy to see that $LD(3)$ does not exist. □

In Lemma 2.2.2, the construction of $LD^*(9)$ is given by Chen Demong [Chen], and the others are given by Lu Jiaxi [Lu 3].

Lemma 2.2.3. There exists an $LD^*(2^n)$ for $n \ge 2$.

Construction. Let $F = GF(2^n)$. Given $a \in F\backslash\{0,1\}$. We are going to construct an $LD(2^n) = (F, \mathcal{L}^1, \mathcal{L}^2, (\mathcal{L}_x)_x)$. Each $\mathcal{L}_x (x \in F)$ consists of the following two types of ordered triples:

$$(x_0, x_1, x_2) \text{ with } x_0, x_1, x_2 \in F\backslash\{x\} \text{ and } x_0 + x_1 + ax_2 = ax; \quad (*)$$
$$(x_1, x_1, x_2) \text{ with } x_1, x_2 \in F\backslash\{x\} \text{ and } x_1 + ax_2 = (1 + a)x. \quad (**)$$

And ordered quadruple systems $\mathcal{L}^1$ and $\mathcal{L}^2$ are defined as follows:

$$\mathcal{L}^1 = \{(x_0, x_1, x_2, x_3) \in F^4 : x_0 + x_2 + x_3 = 0, (a+1)x_0 + x_1 + ax_2 = 0\};$$

$$\mathcal{L}^2 = \{(x_0, x_1, x_2, x_3) \in F^4 : x_1 + x_2 + x_3 = 0, x_0 + (a+1)x_1 + ax_2 = 0\}.$$

Proof. We want to show that all the conditions (C1)—(C6) are satisfied. The condition (C1) is obvious. For any given $u, v \in F$ and any chosen positions i, j, $i \neq j \in \{0, 1, 2, 3\}$, we can uniquely determine a quadruple $(x_0, x_1, x_2, x_3) \in \mathcal{L}^1$ (resp. $\mathcal{L}^2$) such that $x_i = u$ and $x_j = v$. Thus, $\mathcal{L}^1$ (resp. $\mathcal{L}^2$) is a $OA(2, 4, 2^n)$, the condition (C2) holds. But, since the common solution of four conditional equations in $\mathcal{L}^1$ and $\mathcal{L}^2$ is $x_0 = x_1 = x_2$ and $x_3 = 0$, thus (C4) is satisfied. Furthermore, for (C3), we have these facts (given $x \in F$):

(i) For any given $x_0, x_1 \in F\backslash\{x\}$, if $x_0 \neq x_1$ the equation $(*)$ determines a unique $x_2 \in F\backslash\{x\}$ such that $(x_0, x_1, x_2) \in \mathcal{L}_x$; and if $x_0 = x_1$ the equation $(**)$ determines a unique $x_2 \in F\backslash\{x\}$ such that $(x_1, x_1, x_2) \in \mathcal{L}_x$.

(ii) For any given $x_0, x_2 \in F\backslash\{x\}$, if $x_0 + ax_2 \neq (1+a)x$, the equation $(*)$ determines a unique $x_1 \in F\backslash\{x\}$ such that $(x_0, x_1, x_2) \in \mathcal{L}_x$; and if $x_0 + ax_2 = (1 + a)x$, then $(x_0, x_0, x_2) \in \mathcal{L}_x$.

(iii) Similar to (ii), for any given $x_1, x_2 \in F\backslash\{x\}$ there is a unique x_0 such that $(x_0, x_1, x_2) \in \mathcal{L}_x$.

Furthermore, let T be an ordered triple (x_0, x_1, x_2) from F. If $(a+1)x_0+x_1+ax_2 = 0$, then $(x_0, x_1, x_2, x_0+x_2) \in \mathcal{L}^1$; if $x_0+(a+1)x_1+ax_2 = 0$ then $(x_0, x_1, x_2, x_1+x_2) \in \mathcal{L}^2$; or else $T \in \mathcal{L}_x$ where $x = (1+a)^{-1}(x_1+ax_2)$ when $x_0 = x_1$, or $x = a^{-1}(x_0 + x_1 + ax_2)$ when $x_0 \neq x_1$. The condition (C5) is satified.

Lastly, the quadruple system $\mathcal{L}^0 = \{(x_0, x_1, x_2, x_3) \in F^4\text{:}\ \ (a+1)x_0 + x_2 + ax_3 = 0 \text{ and } x_1 + (a+1)x_2 + ax_3 = 0\}$ by the condition equations of $\mathcal{L}^1$ and $\mathcal{L}^2$. It is easy to see that (C6) is satisfied. □

This Lemma (i.e. (L1)) was given by Lu. And, the next Lemma was given in [Kang1], which is an extension of (L6).

2.2.3 Recursive constructions

Lemma 2.2.4. If an $LD(n)$ exists then there exists an $LD(3n)$.

Construction. Let $I_n = \{0, 1, ..., n-1\}$ be a set of n elements ($n > 1$) and $Z_3 = \{0, 1, 2\}$ be the residue class ring modulo 3. Suppose there exists an $LD(n) = (I_n, \hat{\mathcal{L}}^1, \hat{\mathcal{L}}^2, (\hat{\mathcal{L}}_t)_{t \in I_n})$, where $c_0 = 0$ in the condition (C4) satisfied by the design. For convenience, we denote $\hat{\mathcal{L}}^1 = M$ and $\overline{M} = \{(i_0, i_1, i_2):$ there exists $i_3 \in I_n$ such that $(i_0, i_1, i_2, i_3) \in M\}$. We define the following mappings from I_n into I_n:

If $(0, j, x, \alpha)$ and $(i, j, y, \beta) \in M$, let

$$\begin{aligned} g_i &: \ x \longmapsto y \ (g_i \text{ is denoted by } f_1 \text{ or } f_2 \text{ when } \beta = 0 \text{ or } 1), \\ \phi_j &: \ y \longmapsto i, \\ \psi_i &: \ y \longmapsto j, \\ \sigma &: \ i \longmapsto y \ (\text{only for } \beta = 1). \end{aligned}$$

It is easy to see that these mappings are all one-to-one. And we have

$$\begin{cases} f_1(x) = i \text{ if and only if } g_i(x) = i, \\ f_2(x) = \sigma(i) \text{ if and only if } g_i(x) = \sigma(i). \end{cases}$$

Now, we will construct an $LD(3n) = (Z_3 \times I_n, \mathcal{L}^1, \mathcal{L}^2, (\mathcal{L}_{x,t})_{x \in Z_3, t \in I_n})$ as follows (where k runs over Z_3).

Each system $\mathcal{L}^j (j = 1, 2)$ consists of three parts (each part contains $3n^2$ ordered quadruples):

part 1. $((k, i_0), (k, i_1), (k, i_2), (0, i_3))$ provided $(i_0, i_1, i_2, i_3) \in \hat{\mathcal{L}}^j$;

part 2. $((k, i_0), (k+1, i_1), (k-1, f_j(i_2)), (1, i_3))$ provided $(i_0, i_1, i_2, i_3) \in \overline{M}$;

part 3. $((k, i_1), (k-1, i_0), (k+1, f_j(i_2), (2, i_3))$ provieded $(i_0, i_1, i_2, i_3) \in \overline{M}$.

Each system $\mathcal{L}_{0,t} (t \in I_n)$ consists of eight parts:

part 1. $((0, i_0), (0, i_1), (0, i_2))$ provided $(i_0, i_1, i_2) \in \hat{\mathcal{L}}_t$ and $i_0 \neq t \neq i_1$; this gives $(n-1)^2$ ordered triples.

part 2. $((0, i_0), (\pm 1, i_1), (\pm 1, \psi_t(i_2)))$ provided $(i_0, i_1, i_2) \in \overline{M}$ and $i_0 \neq t$; this gives $2n(n-1)$ ordered triples.

part 3. $((\pm 1, i_1), (0, i_0), (\pm 1, \psi_t(i_2)))$ provided $(i_0, i_1, i_2) \in \overline{M}$ and $i_0 \neq t$; this gives $2n(n-1)$ ordered triples.

part 4. $((1,i_0), (2,i_1), (0,g_t(i_2)))$, $((2,i_1), (1,i_0), (0,g_t(i_2)))$ provided $(i_0,i_1,i_2) \in \overline{M}$ and $g_t(i_2) \neq t$, $\sigma(t)$; this gives $2n(n-2)$ ordered triples.

part 5. $((1,i_0),(2,i_1), (1,i_0)),((2,i_1), (1,i_0),(1,i_0))$ with $(i_0,i_1,i_2) \in \overline{M}$ and $g_t(i_2) = t$; this gives $2n$ ordered triples.

part 6. $((1,i_0),(2,i_1), (2,i_1)),((2,i_1), (1,i_0),(2,i_1))$ provided $(i_0,i_1, i_2) \in \overline{M}$ and $g_t(i_2) = \sigma(t)$; this gives $2n$ ordered triples.

part 7. $((1,i_0), (1,i_1), (2,\psi_{i_2}(g_t^{-1}(\sigma(t)))))$, $((2,i_0), (2,i_1), (1,\phi_{i_2}(g_t^{-1}(t))))$ provided $(i_0, i_1,i_2) \in \overline{M}$ and $i_0 \neq i_1$; this gives $2n(n-1)$ ordered triples.

part 8. $((\pm 1,i),(\pm 1,i),(0,\sigma(t)))$ provided $i \in I_n$; this gives $2n$ ordered triples.

And, each system $\mathcal{L}_{x,t}(x \in Z_3\backslash\{0\}, t \in I_n)$ consists of the ordered triples $((k_0+x,i_0), (k_1+x,i_1), (k_2+x,i_2))$ with $((k_0,i_0),(k_1,i_1), (k_2,i_2)) \in \mathcal{L}_{0,t}$.

Proof. The detailed process is tedious (refer to [Kang 1] and [Kang 2]). Here, we will only give the verification for (C5).

For any ordered triple $T = ((k_0,i_0),(k_1,i_1), (k_2,i_2))$ of the set $Z_3 \times I_n$, let us show that there exists an $\mathcal{L}_{x,t}$ $(x \in Z_3,\ t \in I_n)$ or an $\mathcal{L}^j (j = 1,2)$ such that $T \in \mathcal{L}_{x,t}$ or $T \in \bar{\mathcal{L}}^j$.

1. $k_0 = k_1 = k_2 (= k)$

(1) If $(i_0,i_1,i_2) \in \hat{\mathcal{L}}_t$ then $T \in$ part 1 of $\mathcal{L}_{k,t}$.

(2) If $(i_0,i_1,i_2,*) \in \hat{\mathcal{L}}^j$ then $T \in$ part 1 of $\bar{\mathcal{L}}^j$.

2. $k_0 = k_1 \neq k_2$

(1) If $i_0 = i_1$, let $t = \sigma^{-1}(i_2)$ then $T \in$ part 8 of $\mathcal{L}_{k_2,t}$.

(2) If $i_0 \neq i_1$, let $i_2' \in I_n$ such that $(i_0,i_1,i_2') \in \overline{M}$.

i) When $k_2 \equiv k_0 + 1 \pmod 3$, let $t = \sigma^{-1} f_2(\psi_{i_2'}^{-1}(i_2))$, i.e., $\sigma(t) = f_2(\psi_{i_2'}^{-1}(i_2)), = g_t(\psi_{i_2'}^{-1}(i_2))$ then $T \in$ part 7 of $\mathcal{L}_{k_0-1,t}$.

ii) When $k_2 \equiv k_0 - 1 \pmod 3$, let $t = f_1(\phi_{i_2'}^{-1}(i_2))$, i.e., $t = g_t(\phi_{i_2'}^{-1}(i_2))$, then $T \in$ part 7 of $\mathcal{L}_{k_0+1,t}$.

3. $k_0 \equiv k_1 - 1 \pmod 3$

(1) If $k_2 = k_1$, let $i_2' \in I_n$ such that $(i_0,i_1,i_2') \in \overline{M}$.

i) When $i_2 = i_1$, let $t = \sigma^{-1} f_2(i_2')$, i.e. $\sigma(t) = g_t(i_2')$, then $T \in$ part 6 of $\mathcal{L}_{k_0-1,t}$.

ii) When $i_2 \neq i_1$, let $t \in I_n$ such that $\psi_t(i'_2) = i_2$, then $t \neq i_0$ (or else, from $(t, i_2, i'_2) \in \overline{M}, i_2 = i_1$, a contradition) so $T \in$ part 2 of $\mathcal{L}_{k_0,t}$.

(2) If $k_2 \equiv k_1 - 1 \pmod 3$, let $i'_2 \in I_n$ such that $(i_1, i_0, i'_2) \in \overline{M}$.

i) When $i_2 = i_0$, let $i'_2 \in I_n$ such that $(i_0, i_1, i'_2) \in \overline{M}$, let $t = f_1(i'_2)$, i.e. $t = g_t(i'_2)$, then $T \in$ part 5 of $\mathcal{L}_{k_0-1,t}$.

ii) When $i_2 \neq i_0$, let $i'_2, t \in I_n$ such that $(i_1, i_0, i'_2) \in \overline{M}$ and $\psi_t(i'_2) = i_2$ then $t \neq i_1$ (or else, from $(t, i_2, i'_2) \in \overline{M}, i_2 = i_0$, a contradition) so $T \in$ part 3 of $\mathcal{L}_{k_1,t}$.

(3) If $k_2 \equiv k_1 + 1 \pmod 3$, let $i'_2, i_3, t \in I_n$ such that $(i_0, i_1, i'_2, i_3) \in M$ and $g_t(i'_2) = i_2$.

i) When $i_2 = t$ (then $t = g_t(i'_2)$ i.e. $f_1(i'_2) = t = i_2$) then $T \in$ part 2 of $\bar{\mathcal{L}}^1$.

ii)When $i_2 = \sigma(t)$ (then $\sigma(t) = g_t(i'_2)$, i.e. $f_2(i'_2) = \sigma(t) = i_2$) then $T \in$ part 2 of $\bar{\mathcal{L}}^2$.

iii) Or else $T \in$ part 4 of $\mathcal{L}_{k_2,t}$.

4. $k_0 \equiv k_1 + 1 \pmod 3$

(1) If $k_2 = k_1$

i) When $i_2 = i_1$, let $i'_2, t \in I_n$ such that $(i_1, i_0, i'_2) \in \overline{M}$ and $t = f_1(i'_2)$, i.e. $t = g_t(i'_2)$, then $T \in$ part 5 of $\mathcal{L}_{k_0+1,t}$.

ii) When $i_2 \neq i_1$, let $i'_2, t \in I_n$ such that $(i_0, i_1, i'_2) \in \overline{M}$ and $\psi_t(i'_2) = i_2$ then $t \neq i_0$ (see 3(1)) so $T \in$ part 2 of $\mathcal{L}_{k_0,t}$.

(2) If $k_2 \equiv k_1 + 1 \pmod 3$, let $i'_2 \in I_n$ such that $(i_1, i_0, i'_2) \in \overline{M}$.

i) When $i_2 = i_0$, let $t = \sigma^{-1} f_2(i'_2)$, *i.e.* $\sigma(t) = g_t(i'_2)$, then $T \in$ part 6 of $\mathcal{L}_{k_0+1,t}$.

ii) When $i_2 \neq i_1$, let $t \in I_n$ such that $\psi_t(i'_2) = i_2$, then $t \neq i_1$ (see 3(2)), so $T \in$ part 3 of $\mathcal{L}_{k_1,t}$.

(3) If $k_2 \equiv k_1 - 1 \pmod 3$, let $i'_2, i_3, t \in I_n$ such that $(i_0, i_1, i'_2, i_3) \in M$ and $g_t(i'_2) = i_2$.

i) When $i_2 = t$ (then $t = g_t(i'_2)$, i.e. $f_1(i'_2) = t = i_2$) then $T \in$ part 3 of $\bar{\mathcal{L}}^1$.

ii) When $i_2 = \sigma(t)$ (then $\sigma(t) = g_t(i'_2)$, i.e. $f_2(i'_2) = \sigma(t) = i_2$) then $T \in$ part 3 of $\bar{\mathcal{L}}^2$.

iii) Or else $T \in$ part 4 of $\mathcal{L}_{k_2,t}$. □

Lemma 2.2.5. If there exists a $B[K,1;v]$ and there exists an $LD^*(k)$ for each $k \in K$, then there exists an $LD^*(v)$.

Construction. Let there be a $B[K,1;v] = (X, \mathcal{B})$. Given an element $c_0 \in X$. There exists an $LD^*(|B|) = (B, M_B^1, M_B^2, (M_{B,x})_{x \in B})$ for each $B \in \mathcal{B}$. If $c_0 \in B$, let $\mathcal{L}_B^j = M_B^j$, $j = 1,2$, and we can rearrange the elements of B such that $(x,x,x,c_0) \in \mathcal{L}_B^1 \cap \mathcal{L}_B^2$ for any $x \in B$. If $c_0 \notin B$, let $\mathcal{L}_B^j = \{(x_0,x_1,x_2,x_3) : (x_0,x_1,x_2) \in \overline{M}_B^j,\ (x_3,x_1,x_2) \in \overline{M}_B^{3-j}, x_0 \neq x_1\}, j = 1,2$. As $2 \notin K$ (there exists no $LD(2)$), we can construct an idempotent quasigroup $(B, \circ_B)$, for each $B \in \mathcal{B}$. Define

$$\mathcal{L}^j = \bigcup_{B \in \mathcal{B}} \mathcal{L}_B^j,\ j = 1,2;$$

$$\mathcal{L}_x = (\bigcup_{B \in \mathcal{B}} M_{B,x}) \bigcup \{(x_1,x_2,x_3) \in (X \backslash \{x\})^3 : \{x_1,x_2,x_3\} \text{ is not contained in any } B \text{ of } \mathcal{B}, \text{ and } x_1 \circ_{x_1x_2} x_2 = x_3 \circ_{x_3x} x\},\ x \in X,$$

where the symbol $\circ_{xy}$ expresses the operation in the quasigroup corresponding the block $B \in \mathcal{B}$, which contains x and y. Then we obtaine an expected $LD^*(v) = (X, \mathcal{L}^1, \mathcal{L}^2, (\mathcal{L}_x)_{x \in X})$.

Proof. We need to show the conditions (C2)–(C6) are satisfied.

(C2) We discuss only quadruple system $\mathcal{L}^1$, another system $\mathcal{L}^2$ is similar to it. We need to show the "fact" that for any given $u, v \in X$ and any chosen two positions i, j from the set $\{0,1,2,3\}$ there is unique quadruple $(x_0,x_1,x_2,x_3) \in \mathcal{L}^1$ such that $x_i = u$ and $x_j = v$.

If $u = v \neq c_0$, then there is unique block $B \in \mathcal{B}$ such that $\{u, c_0\} \subset B$, thereby, since $\mathcal{L}_B^1 = M_B^1$ is an $OA(2,4,|B|)$, the "fact" holds. If $u = v = c_0$, then there is a block $B \in \mathcal{B}$ with $c_0 \in B$, thereby $(c_0,c_0,c_0,c_0) \in \mathcal{L}_B^1$. Note that in the second case, such B is not unique, but we get only a copy in the union set $\mathcal{L}^1 = \bigcup_{B \in \mathcal{B}} \mathcal{L}_B^1$.

For the remaining case $u \neq v$, there is unique block $B \in \mathcal{B}$ such that $\{x,y\} \subset B$. If $c_0 \in B$ then similar to the first case before, or else for different choose of two positions $i, j \in \{0,1,2,3\}$, we have the following discussion:

(i) $i = 0, j = 1$. There are unique $(u,v,x_2,w) \in M_B^1$ and unique $(x_3,v,x_2,w') \in M_B^2$, thereby $(u,v,x_2,x_3) \in \mathcal{L}_B^1 \subset \mathcal{L}^1$.

(ii) $i = 0, j = 2$. There are unique $(u,x_1,v,w) \in M_B^1$ and unique

$(x_3, x_1, v, w') \in M_B^2$, thereby $(u, x_1, v, x_3) \in \mathcal{L}_B^1 \subset \mathcal{L}^1$.

(iii) $i = 0, j = 3$. Since $LD^*(|B|)$ satisfies the condition (C6), there is unique $(u, v, x_1, x_2) \in M_B^0$, i.e., $(u, x_1, x_2) \in M_B^1$ and $(v, x_1, x_2) \in M_B^2$, thus $(u, x_1, x_2, v) \in \mathcal{L}_B^1 \subset \mathcal{L}^1$.

(iv) $i = 1, j = 2$. There are unique $(x_0, u, v, w) \in M_B^1$ and unique $(x_3, u, v, w') \in M_B^2$, thereby $(x_0, u, v, x_3) \in \mathcal{L}_B^1 \subset \mathcal{L}^1$.

(v) $i = 1, j = 3$. There are unique $(v, u, x_2, w) \in M_B^2$ and unique $(x_0, u, x_2, w') \in M_B^1$, thereby $(x_0, u, x_2, v) \in \mathcal{L}_B^1 \subset \mathcal{L}^1$.

(vi) $i = 2, j = 3$. There are unique $(v, x_1, u, w) \in M_B^2$ and unique $(x_0, x_1, u, w') \in M_B^1$, thereby $(x_0, x_1, u, v) \in \mathcal{L}_B^1 \subset \mathcal{L}^1$.

(C3) Take any $y, z \in X \backslash \{x\}$. If $\{x, y, z\}$ is contained in certain block B of $\mathcal{B}$ (of course, B is unique), then, since $M_{B,x}$ is an $OA(2, 3, |B \backslash \{x\}|)$, the ordered 2-tuple (y, z) occurs exactly in a member of $\mathcal{L}_{B,x}$ when we run two fingers down any two columns. If $\{x, y, z\}$ is contained in none block of $\mathcal{B}$ (then $y \neq z$), there exists unique w_1 (resp. w_2 or w_3)$\in X \backslash \{x\}$ such that $w_1 \circ_{w_1 y} y = z \circ_{zx} x$ (resp. $y \circ_{y w_2} w_2 = z \circ_{zx} x$ or $y \circ_{yz} z = w_1 \circ_{w_1 x} x$). Thereby, there exists unique ordered triple (w_1, y, z) (resp. (y, w_2, z) or (y, z, w_3)) $\in \mathcal{L}_x$.

(C4) For any $x \in X$, if $x \neq c_0$ then there exists unique $B \in \mathcal{B}$ such that $\{x, c_0\} \subset B$, thereby $(x, x, x, c_0) \in M_B^1 \cap M_B^2 = \mathcal{L}_B^1 \cap \mathcal{L}_B^2 \subset \mathcal{L}^1 \cap \mathcal{L}^2$; if $x = c_0$ then there exists block $B \in \mathcal{B}$ thus $(c_0, c_0, c_0, c_0) \in M_B^1 \cap M_B^2 \subset \mathcal{L}_B^1 \cap \mathcal{L}_B^2 \subset \mathcal{L}^1 \cap \mathcal{L}^2$.

(C5) For any ordered triple $T = (x_0, x_1, x_2)$ from X, if $x_0 = x_1 = x_2$ then $(x_0, x_1, x_2, c_0) \in \mathcal{L}^1 \cap \mathcal{L}^2$ by (C4), or else:

(i) If $\{x_0, x_1, x_2\}$ is contained in certain block B of $\mathcal{B}$, then since $LD^*(|B|)$ satisfies (C5), T is either contained in an $M_{B,x}$ ($\subset \mathcal{L}_x$, $x \in B$) or in M_B^j ($\subseteq \mathcal{L}_B^j \subset \mathcal{L}^j$, $j \in \{1, 2\}$).

(ii) If $\{x_0, x_1, x_2\}$ is contained in none block of $\mathcal{B}$, then there is unique $x \in X \backslash \{x_0, x_1, x_2\}$ such that $x_0 \circ_{x_0 x_1} x_1 = x_2 \circ_{x_2 x} x$, thus $T \subset \mathcal{L}_x$.

(C6) For any ordered pair $P = (x, y) \in X^2$, we only need to show that there is an ordered quadruple $(x, y, s, t) \in \mathcal{L}^0$, where $\mathcal{L}^0 = \{(x_0, x_1, x_2, x_3)$: $(x_0, x_2, x_3) \in \overline{\mathcal{L}}^1$ and $(x_1, x_2, x_3) \in \overline{\mathcal{L}}^2\}$. If $x = y$, we have $(x, x, x) \in \overline{\mathcal{L}}^1$ $\cap \overline{\mathcal{L}}^2$ by (C4), thereby $(x, x, x, x) \in \mathcal{L}^0$. (Other cases are similar. Only need point that if $(x,y,z,x) \in \mathcal{L}^0$, then $x = y = z$.) If $x \neq y$, then there is a unique $B \in \mathcal{B}$ such that $\{x,y\} \subset B$. When $c_0 \in B$, the conclusion holds since $LD^*(|B|)$ satisfies (C6). And when $c_0 \notin B$, there exists unique $(x,y,s,t) \in M_B^0$, i.e. $(x,s,t) \in \overline{M}_B^1$ and $(y,s,t) \in \overline{M}_B^2$. By the definition of $\mathcal{L}_B^j$, $(x,s,t,y) \in \mathcal{L}_B^1$ and $(y,s,t,x) \in \mathcal{L}_B^2$, thus $(x,y,s,t) \in \mathcal{L}_B^0 \subset \mathcal{L}^0$.□

2.2.4 Corollaries and conclusion

Corallary 2.2.6. There exists an $LD^*(v)$ for $v \equiv 1,4$ (and 12) and $v > 1$.

Proof. It is well known that there exists a $B[4,1;v]$ for $v \equiv 1,4$ (mod 12) and $v > 1$ [BJL]. And, there exists an $LD^*(4)$ by Lemma 2.2.2. Thus, there exists an $LD^*(v)$ by Lemma 2.2.5. □

Corollary 2.2.7. There exists an $LD^*(v)$ for $v \equiv 7,10$ (mod 12) and $v \neq 10$.

Proof. By [Brou 1], there exists a $B[\{4,7\},1;v]$ for $v \equiv 7,10$ (mod 12) and $v \neq 10, 19$. And, there exist an $LD^*(4)$ by Lemma 2.2.2 and an $LD^*(7)$ by Lemma 2.2.3. Thus, there exists an $LD^*(v)$ by Lemma 2.2.5. As for $LD^*(19)$, we can get it immediately from Lemma 2.2.3. □

Corollary 2.2.8. For $n \equiv k+4$ (mod 12), $n \geq n_0 = 12\lceil\frac{k-1}{4}\rceil + k + 4$, if there exists an $LD^*(k)$ or $k = 1$ then there exists an $LD^*(n)$.

Proof. By [BJL], there exists a resolvable $B[4,1;v]$ for $v \equiv 4$ (mod 12). Thus, there exists a $RB[4,1;n-k] = (X,\mathcal{B})$. Here, the number of parallel classes in $\mathcal{B}$ is

$$\frac{n-k-1}{4-1} \geq 4\lceil\frac{k-1}{4}\rceil + 1 \geq k.$$

Choosing any k parallel classes from $\mathcal{B}$, γ_1, γ_2, ..., γ_k, we can obtain a $B[\{4,5,k\},\ 1;n] = (X \bigcup \{\infty_1,\ ...,\infty_k\},\ \mathcal{A})$, where $\infty_1, \infty_2,\ ..., \infty_k \notin X$ and $\mathcal{A}$ contains the following blocks:

$$\begin{cases} \{\infty_1, \infty_2, ..., \infty_k\}; \\ B \in (\mathcal{B} \setminus \bigcup_{i=1}^{k} \gamma_i); \\ B \bigcup \{\infty_i\} \text{ with } B \in \gamma_i,\ 1 \leq i \leq k. \end{cases}$$

Therefore, there exists an $LD^*(n)$ by Lemma 2.2.5 and the existence of $LD^*(4)$, $LD^*(5)$ and $LD^*(k)$. □

Corollary 2.2.9. There exists an $LD^*(54)$.

Proof. First, there exists a $RB[q,1;q^n]$ for any prime power q and any positive integer n [BJL]. Taking $q=7$ and $n=2$, we have a $RB[7,1;49]$, which contains eight parallel classes. Follow the last Collorally, adding five new elements to the design, we can obtain a $B[\{5,7,8\},1;49+5]$, Thus, an $LD^*(54)$ is obtained by Lemma 2.2.5 and the existence of $LD^*(5)$, $LD^*(7)$ and $LD^*(2^3)$. □

Corollary 2.2.10. Let $v=kn+s$, $1\le s\le n+1$. If there exist $LD^*(k), LD^*(k+1), LD^*(n+1), LD^*(s)$ and $TD(k+1,n)$, then there exists an $LD^*(v)$.

Proof. It is easy to know that the existence of $TD(k+1,n)$ is equivalent to the existence of a resolvable $TD(k,n)$. Let n parallel classes of the resolvable $TD(k,n)=(X,\mathcal{G},\mathcal{A})$ be $\mathcal{A}_1,\mathcal{A}_2,...,\mathcal{A}_n$. Adding s elements $\infty_0,\infty_1,...,\infty_{s-1}$ and defining

$$\begin{aligned}\mathcal{A}_i' &= \{B\bigcup\{\infty_i\}:\ B\in\mathcal{A}_i\}, 1\le i\le s-1,\\ \mathcal{G}' &= \{G\bigcup\{\infty_0\}:\ G\in\mathcal{G}\},\end{aligned}$$

we can obtain a $B[\{n+1,k+1,k,s\},1;v]$ on the set $X\bigcup\{\infty_0,...,\infty_{s-1}\}$, the block set of which is

$$\{G':G'\in\mathcal{G}'\}\bigcup\{B':B'\in\bigcup_{i=1}^{s-1}\mathcal{A}_i'\}\bigcup\{B:B\in\bigcup_{j=s}^{n}\mathcal{A}_j\}\bigcup\{\infty_0,\infty_1,...,\infty_{s-1}\}.$$

Therefore, by Lemma 2.2.5 and the existence of $LD^*(n+1), LD^*(k+1), LD^*(k)$ and $LD^*(s)$, there exists an $LD^*(v)$. □

Corollary 2.2.11. For $v=kn+s, s\le n$, if there exist $LD^*(n)$, $LD^*(s)$, $LD^*(k)$ and $LD^*(k+1)$, and there exists a $TD(k+1,n)$, then there exists an $LD^*(v)$.

Proof. Let $TD(k+1,n)=(X,\mathcal{G},\mathcal{A})$, where $\mathcal{G}$ is a partition of X into $k+1$ n-groups $G_1,G_2,...,G_{k+1}$. Choose a $(n-s)$-subset T from G_{k+1}. Define a collection $\mathcal{B}$ of blocks as follows:

$$\mathcal{B}=\{B-T;B\in\mathcal{A}\}\bigcup\{G_i;1\le i\le k\}\bigcup(G_{k+1}\backslash T).$$

Then $(X\backslash T, \mathcal{B})$ is a $B[\{k, k+1, n, s\}, 1; nk+s]$. Therefore, there exists an $LD^*(nk+s)$ by Lemma 2.2.5. Note that the existence condition for $LD^*(s)$ is not necessary, when $s = 0, 1$ or n. □

Corollary 2.2.12. For $v = 4n + s$, $s \leq n$ and $n \neq 10$, if there exist $LD^*(n)$ and $LD^*(s)$ then there exists an $LD^*(v)$.

Proof. Take $k = 4$ in Corollary 2.2.10. Since there exist three pairwise orthogonal Latin squares of order n for $n \neq 2, 3, 6, 10$[BJL], thus there exists a $TD(5, n)$. But, there are not $LD^*(2)$, $LD^*(3)$, and $LD^*(6)$ by Lemma 2.2.3. Therefore the Corollary 2.2.10 can be used. □

Now, by Lemmas 2.2.2—2.2.5 and Corollaries 2.2.6—2.2.12, we can prove Theorem 2.2.1.

Proof of Theorem 2.2.1

For all integers $v \geq 4, v \neq 6$, we discuss respectively the following cases according to the residue of v (mod 12).

(1) $v \equiv 1, 4, 7, 10 \pmod{12}$. By Corollary 2.2.6 and 2.2.7, the unknown design is only $LD(10)$.

(2) $v \equiv 5 \pmod{12}$. By taking $k = 1$ ($n_0 = 5$) in Corollary 2.2.8, we can obtain all $LD^*(v)$ for such orders v.

(3) $v \equiv 8 \pmod{12}$. By taking $k = 2^2$ ($n_0 = 20$) in Corollary 2.2.8, the existence of $LD^*(v)$ for $v \geq 20$ can be obtained. And there exists an $LD^*(8)$ by Lemma 2.2.2.

(4) $v \equiv 9 \pmod{12}$. By taking $k = 5$ ($n_0 = 21$) in Corollary 2.2.8, we can obtain all $LD^*(v)$ for $v \equiv 9 \pmod{12}$ except $v = 9$. And, an $LD^*(9)$ is given by Lemma 2.2.3.

(5) $v \equiv 11 \pmod{12}$. By taking $k = 7$ ($n_0 = 35$) in Corollary 2.2.8, the existence of $LD^*(v)$ for $v \geq 35$ can be obtained. And, there exists an $LD^*(11)$ by Lemma 2.2.3. Thus, the remaining unknown design is only $LD(23)$.

(6) $v \equiv 0 \pmod{12}$. By taking $k = 2^3$ ($n_0 = 36$) in Corollary 2.2.8, we can obtain $LD^*(v)$ for $v \geq 36$. And, there exists an $LD^*(24)$ by Corollary 2.2.12 (taking $n = 5, s = 4$). But, $LD(4) \rightarrow LD(12)$ by Lemma 2.2.4, thus the remaining is only $LD^*(12)$.

(7) $v \equiv 2 \pmod{12}$. By taking $k = 22$ ($n_0 = 98$, $LD^*(22)$ has been

gotten from case (1)) in Corollary 2.2.8, the existence of $LD^*(v)$ for $v \geq 98$ can be obtained. And, there exist an $LD^*(50)$ (taking $k = n = 7, s = 1$) and an $LD^*(86)$ (taking $k = 7, n = 11, s = 9$) by Corollary 2.2.11 . The remaining orders are 14, 26, 38, 62 and 74 (for LD).

(8) $v \equiv 3 \pmod{12}$. By taking $k = 11 (n_0 = 51)$ in Corollary 2.2.8, the existence of $LD^*(v)$ for $v \geq 51$ can be obtained. And, there exists an $LD^*(39)$ by Corollary 2.2.12 (taking $n = 8, s = 7$). But $LD(5) \to LD(15)$ and $LD(9) \to LD(27)$ by Lemma 2.2.4. Thereby, the remaining unknown design are $LD^*(15)$ and $LD^*(27)$.

(9) $v \equiv 6 \pmod{12}$. By taking $k = 50$ ($n_0 = 210$) in Corollary 2.2.8, we can obtain $LD^*(v)$ for $v \geq 210$. An $LD^*(114)$ can be obtained by Corollary 2.2.10, since 114=4×23+22 and there exist $LD^*(4)$, $LD^*(5)$, $LD^*(24)$, $LD^*(22)$ and $TD(5, 23)$. Furthermore, there exist $LD^*(v)$ for the following orders v: 54 (by Corollary 2.2.9); $78 = 7 \times 11 + 1$, $102 = 7 \times 13 + 11$, $126 = 7 \times 17 + 7$ (by Corollary 2.2.11); $138 = 4 \times 29 + 22$, $150 = 4 \times 2^5 + 22$, $162 = 4 \times 35 + 22$, $174 = 4 \times 35 + 34$, $186 = 4 \times 41 + 22$, $198 = 4 \times 44 + 22$ (by Corollary 2.2.12); where the existence of those LD^* relating to Corollary's conditions have been obtained before. Lastly, $LD(22) \to LD(66)$ by Lemma 2.2.4. Therefore, the remaining unknown designs are $LD^*(66)$, $LD(18), LD(30)$, $LD(42)$ and $LD(90)$.

This completes the proof. □

2.3 Lu's contribution

In this section we will discuss a few types of auxiliary designs introduced by Lu Jiaxi and their important role to construct $LSTS$. They are LD-design, which has been studied detailed in last section, AD-design, AD^*-design, AD^{**}-design and LAD, LAD_1, LAD_2, LAD_3.

2.3.1 LD-design and $LSTS$

The following theorem gives an important relation between $LD(n)$ and $LSTS(3n)$.

Theorem 2.3.1. If there exist both an $LD(n)$ and an $LSTS(n+2)$, then there exists an $LSTS(3n)$ also.

Construction. Let

$LD(n) = (I_n, \mathcal{L}^1, \mathcal{L}^2, (\mathcal{L}_i)_{i \in I_n})$ with $(x, x, x, 0) \in \mathcal{L}^1 \bigcap \mathcal{L}^2$, $\forall x \in I_n$;

$LSTS(n+2) = \{(\{\infty_1, \infty_2\} \bigcup I_n, \mathcal{B}_i) : i \in I_n\}$ with $\{\infty_1, \infty_2, i\} \in \mathcal{B}_i$.

From $\mathcal{L}_i$ and $\mathcal{B}_i$ we can construct n triple systems $\mathcal{A}_i$ on the set $Z_3 \times I_n$, $i \in I_n$. Each $\mathcal{A}_i$ consists of the following four parts:

$M_1 : \{(0, i), (1, i), (2, i)\}$;

$M_2 : \{(0, u), (1, v), (2, w)\}$ with $(u, v, w) \in \mathcal{L}_i$;

$M_3 : \{(k, u), (k, v), (k, w)\}$ with $k \in Z_3, u, v, w \in I_n$ and $\{u, v, w\} \in \mathcal{B}_i$;

$M_4 : \{(k, i), (k+j, v), (k+j, w)\}$ with $k \in Z_3, j \in \{1, 2\}, v, w \in I_n \backslash \{i\}$ and $\{\infty_j, v, w\} \in \mathcal{B}_i$.

For $i \in I_n \backslash \{0\}$ and $j \in \{1, 2\}$ we can also construct $2(n-1)$ triple systems $\mathcal{B}_{ij}$ on the set $Z_3 \times I_n$. Each $\mathcal{B}_{ij}$ consists of the following two parts:

$N_1 : \{(0, u), (1, v), (2, w)\}$ with $(u, v, w, i) \in \mathcal{L}^j$;

$N_2 : \{(k, u), (k+j, v), (k+j, w)\}$ provided $k \in Z_3, j \in \{1, 2\}$ and there exists $t \in I_n$ such that $\{\infty_j, v, w\} \in \mathcal{B}_t, (x_0, x_1, x_2, i) \in \mathcal{L}^j$ and $x_k = u, x_{k+j} = t$ where $v, w \in I_n \backslash \{t\}$.

Then $\{(Z_3 \times I_n, \mathcal{A}_i) : i \in I_n\} \bigcup \{(Z_3 \times I_n, \mathcal{B}_{ij}) : i \in I_n \backslash \{0\}, j = 1, 2\}$ will be an expected $LSTS(3n)$.

Proof. Step 1. Each $\mathcal{A}_i$ is an $STS(3n)$, for $i \in I_n$.

Direct calculation shows

$$\begin{aligned} |\mathcal{A}_i| &= |M_1| + |M_2| + |M_3| + |M_4| \\ &= 1 + (n-1)^2 + 3\left[\tfrac{(n+2)(n+1)}{6} - n\right] + 3 \cdot 2 \cdot \tfrac{n-1}{2} \\ &= \tfrac{3n(3n-1)}{6}. \end{aligned}$$

Therefore we only have to show that every pair P of distinct elements of $Z_3 \times I_n$ is contained in some triple of $\mathcal{A}_i$. All the possibilities are exhausted as follows:

(i) $P = \{(k, v), (k, w)\}$, $k \in Z_3$ and $v \neq w \in I_n$. Since $(\{\infty_1, \infty_2\} \bigcup I_n, \mathcal{B}_i)$ is an $STS(n+2)$, there is an element $u \in (\{\infty_1, \infty_2\} \bigcup I_n)$ such that $\{u, v, w\} \in \mathcal{B}_i$. If $u = \infty_j$ then P is contained in some triple of M_4, or else P is contained in certain triple of M_3.

(ii) $P = \{(k, i), (k+j, v)\}, k \in Z_3, j \in \{1, 2\}$ and $v \in I_n$. If $v = i$ then

P is contained in some triple of M_1, or else there is $w \in I_n \backslash \{i, v\}$ such that $\{\infty_j, v, w\} \in \mathcal{B}_i$, so P is contained in some triple of M_4.

(iii) $P = \{(k, u), (k+j, v)\}, k \in Z_3, j \in \{1, 2\}$ and $u, v \in I_n \backslash \{i\}$. Since $\mathcal{L}_i$ satisfies the condition (C3) of $LD(n)$, there exists $w \in I_n \backslash \{i\}$ such that $(u, v, w) \in \mathcal{L}_i$ (when $k = 0, j = 1$) or $(u, w, v) \in \mathcal{L}_i$ (when $k = 0, j = 2$) or $(w, u, v) \in \mathcal{L}_i$ (when $k = 1, j = 1$) or so on. Thus P is contained in some triple of M_2.

Step 2. Each $\mathcal{B}_{ij}$ is an $STS(3n)$ for $i \in I_n \backslash \{0\}, j \in \{1, 2\}$.

Direct calculation shows

$$|\mathcal{B}_{ij}| = |N_1| + |N_2| = n + 3 \cdot n \cdot \frac{n-1}{2} = \frac{3n(3n-1)}{6},$$

therefore we only need to show that every pair P is contained in some triple of $\mathcal{B}_{ij}$.

(i) $P = \{(k, v), (k, w)\}$, $k \in Z_3$ and $v \neq w \in I_n$. Since $\{(\{\infty_1, \infty_2\} \bigcup I_n, \mathcal{B}_i):\ i \in I_n\}$ is an $LSTS(n+2)$, hence there exists $t \in I_n$ such that $\{\infty_j, v, w\} \in \mathcal{B}_t$, $(x_0, x_1, x_2, i) \in \mathcal{L}^j$ and $x_k = t$. It follows that $P \subset \{(k-j, u), (k, v), (k, w)\} \in N_2$, where $u = x_{k-j}$.

(ii) $P = \{(k, u), (k+j, v)\}, k \in Z_3$ and $u, v \in I_n$. Since $\mathcal{L}^j$ satisfies the condition (C2) of $LD(n)$, there exists $(x_0,\ x_1,\ x_2,\ i) \in \mathcal{L}^j$ such that $x_k = u$. Let $x_{k+j} = t$ and $x_{k-j} = w$. If $t = v$ then $P \subset \{(k, u),\ (k+j, v),\ (k-j,\ w)\} \in N_1$, or else there exists $w' \in I_n \backslash \{v\}$ such that $\{\infty_j,\ v, w'\} \in \mathcal{B}_t$, hence $P \subset \{(k, u),\ (k+j, v),\ (k+j,\ w')\} \in N_2$.

Step 3. $\{(Z_3 \times I_n, \mathcal{A}_i) : i \in I_n\} \bigcup \{(Z_3 \times I_n, \mathcal{B}_{ij}) :\ i \in I_n \backslash \{0\}, j \in \{1, 2\}\}$ is an $LSTS(3n)$.

We only have to show that every triple T of distinct elements of $Z_3 \times I_n$ belongs to some $\mathcal{A}_i$ or $\mathcal{B}_{ij}$. All the posibilities are exhausted as follows:

(i) $T = \{(k, u), (k, v), (k, w)\}, k \in Z_3, u, v, w \in I_n$ and $u \neq v \neq w \neq u$. Since $\{(\{\infty_1, \infty_2\} \bigcup I_n, \mathcal{B}_i) :\ i \in I_n\}$ is an $LSTS(n+2)$, hence there exists $i \in I_n$ such that $\{u, v, w\} \in \mathcal{B}_i$. Therefore T belongs to M_3 in $\mathcal{A}_i$.

(ii) $T = \{(0, u), (1, v), (2, w)\}, u, v, w \in I_n$. If $u = v = w$, then T belongs to M_1 in $\mathcal{A}_u$, or else, since the known $LD(n)$ satisfies the condition (C5), the ordered triple (u, v, w) is either contained in an $\mathcal{L}_i (i \in I_n)$ or in an $\overline{\mathcal{L}^j} (j \in \{1, 2\})$, then T belongs to M_1 of $\mathcal{A}_i$ or belongs to N_1 of $\mathcal{B}_{ij}$, where the latter i satisfies $(u, v, w, i) \in \mathcal{L}^j$ and $i \neq 0$.

(iii) $T = \{(k,u),(k+j,v),(k+j,w)\}, k,j \in Z_3, u,v,w \in I_n$ and $v \neq w$. There exists $t \in I_n \backslash \{v,w\}$ such that $\{\infty_j, v, w\} \in \mathcal{B}_t$. If $t = u$ then T belongs to M_4 of $\mathcal{A}_u$, or else there exists $i \in I_n \backslash \{0\}$ such that $(x_0, x_1, x_2, i) \in \mathcal{L}^j$ and $x_k = u, x_{k+j} = t$, hence T belongs to N_2 of $\mathcal{B}_{ij}$. □

2.3.2 *AD*-design and *LSTS*

An *AD-design* is a triad $(\rho^1, \rho^2, \mathcal{J})$, where ρ^1 and ρ^2 each is a collection of ordered pairs of I_{12} and $\mathcal{J}$ is a collection of unordered triples from I_{12} such that all of the following conditions are satisfied:

(A1) $\rho^1 \bigcap \rho^2 = \phi$ and $(i,j) \in \rho^1 \bigcup \rho^2$ implies $(j,i) \notin \rho^1 \bigcup \rho^2$;

(A2) For any $i \in I_{12}$ there exists a unique $j \in I_{12}$ such that either $(i,j) \in \rho^1 \bigcup \rho^2$ or $(j,i) \in \rho^1$;

(A3) Each 2-subset of I_{12} is contained in exactly one member of $\mathcal{J} \bigcup \overline{\rho}^1 \bigcup \overline{\rho}^2$, where $\overline{\rho}^k = \{\{i,j\}: \ (i,j) \in \rho^k\}$, $k = 1,2$.

Let $|\rho^1| = m_1, |\rho^2| = m_2$, then $2m_1 + m_2 = 12$ and $|\mathcal{B}| = 22 - \frac{m_1+m_2}{3}$.

An *LAD* consists of twelve *AD*-designs, $AD_i = (\rho_i^1, \rho_i^2, \mathcal{J}_i), 0 \leq i \leq 11$, which satisfied the additional conditions:

(A4) $\bigcup_{i=0}^{11} \mathcal{J}_i$ is a partition of all triples from I_{12};

(A5) $\rho_i^1 \neq \phi$ if and only if $i \in \{0,1,11\}$. $\rho_0^1 \bigcap \rho_1^1 = \phi$, $\rho_0^1 \bigcup \rho_1^1 = \rho_{11}^1$ and $\rho_{11}^2 = \phi$.

(A6) $(\bigcup_{i=0}^{10} \rho_i^2) \bigcup \{(i,j)$ and (j,i): $\ (i,j) \in \rho_{11}^1\}$ is a partition of all ordered pairs of distinct elements of I_{12}.

The *AD*-design and *LAD* had been introduced by Lu Jiaxi in [Lu 2] to study the existence of *LSTS*. And, he gave the following $LAD = \{AD_t : \ 0 \leq t \leq 11\}$, where $AD_t = (\rho_t^1, \rho_t^2, \mathcal{J}_t)$, $0 \leq t \leq 11$. In the Table 2.1, the entry (j) or j at the intersection the i-th row and the t-th column represents the ordered pair $(i,j) \in \rho_t^1$ or ρ_t^2, respectively. And in the Table 2.2, the entry "i,j" at the intersection the i-th row and the t-th column represents the triple $\{i,j,k\} \in \mathcal{J}_t$. To list these tables in the narrow space we have to replace 10 and 11 by u and v, respectively.

Table 2.1 ρ_t^1, ρ_t^2 in AD_t

	0	1	2	3	4	5	6	7	8	9	u	v
0		9	1	2	3	8	7	6	4	5	u	
1	(2)	7	3	8	v	0	u	4	5	6	9	(2)
2		u	3	8	v	0	9	5	6	4	7	
3	(8)	9	v	0	1	2	7	6	4	5	u	(8)
4	0	(9)	2	3	v	1	5	7	6	u	8	(9)
5	8	(7)	0	1	3	v	6	u	4	9	2	(7)
6	2	(u)	8	v	1	3	4	9	5	7	0	(u)
7	0		2	3	v	1	6	u	4	9	8	
8		7	v	0	1	2	u	4	5	6	9	
9	2		8	v	1	3	5	7	6	u	0	
u	8		0	1	3	v	4	9	5	7	2	
v	(0)	u	1	2	3	8	9	5	6	4	7	(0)

Table 2.2 J_t in AD_t

	0	1	2	3	4	5	6	7	8	9	u	v
0	1,3 2,8 5,9 6,u	1,6 2,7 3,4 5,v 8,u	2,v 3,8 4,9 6,7	1,v 4,7 5,6 9,u	1,2 4,6 5,u 7,9 8,v	3,v 4,5 6,9 7,u	1,4 2,u 3,5 6,v 8,9	1,7 2,4 3,u 5,8 9,v	1,u 2,5 3,9 6,8 7,v	1,9 2,6 3,7 4,8 u,v	1,8 2,3 4,v 5,7	1,5 2,9 3,6 4,u 7,8
1	4,7 5,6 9,u 8,v	2,4 3,u 5,8 9,v	2,8 4,6 5,u 7,9	2,3 4,9 6,7	4,u 5,7	2,v 3,8 5,9 6,u	2,5 3,9 6,8 7,v	2,u 3,5 6,v 8,9	2,9 3,6 4,v 7,8	2,7 3,4 5,v 8,u	2,6 3,7 4,5 u,v	3,v 4,8 6,9 7,u
2	3,v 4,u 5,7	3,5 6,v 8,9	5,9 6,u	4,5 6,9 7,u	3,8 4,7 5,6 9,u	4,6 5,u 7,9	3,6 4,v 7,8	3,9 6,8 7,v	3,7 4,8 u,v	3,u 5,8 9,v	4,9 8,v	3,4 5,v 6,7 8,u
3	4,5 6,9 7,u	6,8 7,v	4,7 5,6 9,u	5,9 6,u 8,v	4,9 6,7	4,u 5,7	4,8 u,v	4,v 7,8	5,v 8,u	6,v 8,9	4,6 5,8 9,v	5,u 7,9
4	6,v 8,9	5,u 6,7 8,v	5,v 8,u	6,8 u,v	5,8	7,8 9,v	7,9	5,9 6,u	9,u	5,7 6,9	7,u	5,6 7,v
5	u,v	6,9	7,8	8,u 7,v	9,v	6,8	7,u 8,v	6,7	7,9	6,u	6,v 9,u	8,9
6	7,8		9,v		8,9 u,v	7,v	9,u		7,u		7,9 8,u	8,v
7	9,v	9,u	u,v	8,9	8,u					8,v		
8						9,u		u,v	9,v			
9												u,v

It is easy to verify that this example satisfies all conditions (A1)—(A6), indeed. The role of LAD is manifested in the following theorem.

Theorem 2.3.2. For odd integer n, if there exist an $LSTS(1+2n)$ and a $TD(14,n)$ then there exists an $LSTS(1+12n)$.

Construction. Let $LSTS(1+2n) = \{\{\infty\} \bigcup (I_2 \times Z_n), \mathcal{B}_{is}) : (i,s) \in (i_2 \times Z_n) \backslash \{(1,0)\}\}$;

$LAD = \{AD_k : k \in I_{12}\}$, where $AD_k = (\rho_k^1, \rho_k^2, \mathcal{J}_k)$ is listed above;

$TD(14,n) = (I_{14} \times Z_n, \mathcal{G}, \mathcal{A})$, where $\mathcal{G} = \{\{i\} \times Z_n : i \in I_{14}\}$.

Here $I_k = \{0,1,...,k-1\}$ for $k \in \{2,11,12,14\}$ and $Z_n = \{0,1,..., n-1\}$. Note that 2 is invertible due to n odd. In what follows, we shall use an expression $\{(i_1,j_1), (i_2,j_2), ..., (i_r,j_r)\} \in \overline{\mathcal{A}}$ to denote the fact that the subset $\{(i_1,j_1), (i_2,j_2), ..., (i_r,j_r)\}$ of $I_{14} \times I_n$ is contained in some block of $\mathcal{A}$. Let $\pi = (0,1,2,...,n-1)$ be a cycle permutation on Z_n. For ordered pair (x,y) of distinct elements of I_{12}, define a mapping f_{xy} : $\{0,1\} \to \{x,y\}$ as $f_{xy}(0) = x$ and $f_{xy}(1) = y$. Write $LSTS(13) = \{(\{\infty\} \bigcup I_{12}, C_k); k \in I_{11}\}$, whose existence has been given by Denniston [Den 2]. Here, C_0 consists of the following tripes:

$\{0,1,6\}, \{0,2,3\}, \{3,4,7\}, \{1,4,5\}, \{5,6,8\}, \{2,6,7\}, \{2,8,9\}, \{0,9,10\}, \{2,4,10\}, \{3,5,10\}, \{0,5,7\}, \{1,7,9\}, \{1,8,10\}, \{3,6,9\}, \{0,4,8\}, \{11,7,8\}, \{11,1,3\}, \{11,2,5\}, \{11,6,10\}, \{11,4,9\}, \{\infty,0,11\}, \{\infty,1,2\}, \{\infty,3,8\}, \{\infty,4,6\}, \{\infty,5,9\}, \{\infty,7,10\}$,

and $C_k = \{\{\sigma^k(x), \sigma^k(y), \sigma^k(z)\}: \{x,y,z\} \in C_0\}$, where permutation $\sigma = (0,1,...,10)$. It can be verified that $(i,j) \in \rho_k^1$ implies $\{\infty, i, j\} \in C_k$ for each $k \in I_{11}$, where ρ_k^1 is from $AD_k = (\rho_k^1, \rho_k^2, \mathcal{J}_k)$, $k \in I_{12} \backslash \{11\}$.

Construct $11n$ triple systems $\mathcal{D}_{ks}$ on the set $X = \{\infty\} \bigcup (I_{12} \times Z_n)$, $k \in I_{11}$ and $s \in Z_n$. Each $\mathcal{D}_{ks}$ consists of the following six parts.

$D_1 : \{\infty, (f_{xy}(i), u), (f_{xy}(j), v)\}$ with $(x,y) \in \rho_k^1$ and $\{\infty, (i,u), (j,v)\} \in \mathcal{B}_{os}$;

$D_2 : \{(f_{xy}(i), u), (f_{xy}(j), v), (f_{xy}(k), w)\}$ with $(x,y) \in \rho_k^1$ and $\{(i,u), (j,v), (k,w)\} \in \mathcal{B}_{os}$;

$D_3 : \{(x,u), (x,v), (y,w)\}$ with $(x,y) \in \rho_k^2$ $u \neq v$ and $\{(12,s), (y,w), (x, \frac{u+v}{2})\} \in \overline{\mathcal{A}}$;

D_4 : $\{(x,u),(y,v),(z,w)\}$ with $\{x,y,z\} \in \mathcal{J}_k$, $x < y < z$, $\{(x,u),(y,v),(12,j),(13,w')\} \in \overline{\mathcal{A}}$ and $\{(z,w),(12,s),(13,w')\} \in \overline{\mathcal{A}}$, where $j \neq s$;

D_5 : $\{\infty,(x,u),(y,v)\}$ with $\{\infty,x,y\} \in C_k$, $\{(x,u),(y,v),(12,s)\} \in \overline{\mathcal{A}}$ and $(x,y),(y,x) \notin \rho_k^1$;

D_6 : $\{(x,u),(y,v),(z,w)\}$ with $\{x,y,z\} \in C_k$ and $\{(x,u),(y,v),(z,w),(12,s)\} \in \overline{\mathcal{A}}$.

Finally, construct $n-1$ triple system $\mathcal{E}_s$ on the set X, $s \in Z_n \backslash \{0\}$. Each $\mathcal{E}_s$ consists of the following three parts:

E_1 : $\{\infty,(f_{xy}(i),u),(f_{xy}(j),v)\}$ with $(x,y) \in \rho_{11}^1$ and $\{\infty,(i,u),(j,v)\} \in \mathcal{B}_{1s}$;

E_2 : $\{(f_{xy}(i),u),(f_{xy}(j),v),(f_{xy}(k),w)\}$ with $(x,y) \in \rho_{11}^1$ and $\{(i,u),(j,v),(k,w)\} \in \mathcal{B}_{1s}$;

E_3 : $\{(x,u),(y,v),(z,\pi^s(w))\}$ with $\{x,y,z\} \in \mathcal{J}_{11}$, $x < y < z$ and $\{(x,u),(y,v),(z,w)\} \in \overline{\mathcal{A}}$.

Then $\{(X,\mathcal{D}_{ks}) : k \in I_{11}, s \in Z_n\} \bigcup \{(X,\mathcal{E}_s) : s \in Z_n \backslash \{0\}\}$ will be an expected $LSTS(1+12n)$.

Proof. Step 1. Each $\mathcal{D}_{ks}$ is an $STS(1+12n)$.

Direct calculation shows

$$|\mathcal{D}_{ks}| = |D_1 \bigcup D_2| + \sum_{i=3}^{6} |D_i|$$
$$= \tfrac{2n(2n+1)}{6} m_1 + \tfrac{n(n-1)}{2} m_2 + (22 - \tfrac{m_1+m_2}{3})\,(n-1)n + (6-m_1)n + 20n$$
$$= \tfrac{12n(12n+1)}{6},$$

where $m_1 = |\rho_k^1|$ and $m_2 = |\rho_k^2|$. Therefore, we only have to show that every pair P of distinct elements of X is contained in some triple in $\mathcal{D}_{ks}$. All the posibilities are exhausted as follows:

(i) $P = \{\infty,(x,u)\}$. If there exists $y \in I_{12}$ such that (x,y) or $(y,x) \in \rho_k^1$ then P is contained in D_1, or else there is $y' \in I_{12}$ such that $\{\infty,x,y'\} \in C_k$ hence P is contained in D_5.

(ii) $P = \{(x,u),(x,v)\}, u \neq v$. If there exists $y \in I_{12}$ such that (x,y) or $(y,x) \in \rho_k^1$ then P is contained in $D_1 \bigcup D_2$, or else there exists $y' \in I_{12}$ by (A2) such that $(x,y') \in \rho_k^2$ hence P is contained in D_3.

(iii) $P = \{(x,u),(y,v)\}, x < y$. If (x,y) or $(y,x) \in \rho_k^1$ then P is contained in $D_1 \bigcup D_2$. Or else consider three cases:

(iiia) $\{(x,u),(y,v),(12,s)\} \in \overline{\mathcal{A}}$. There exists $z \in \{\infty\} \bigcup I_{12}$ such that $\{x, y, z\} \in C_k$. If $z = \infty$ then P is contained in D_5, or else P is contained in D_6.

(iiib) $\{(x, u), (y, v), (12, s)\} \notin \overline{\mathcal{A}}$ and $(x, y) \in \rho_k^2$ (similar for $(y, x) \in \rho_k^2$). There exists $j \in Z_n \backslash \{u\}$ such that $\{(x, j), (y, v), (12, s)\} \in \overline{\mathcal{A}}$ thus $P \subset \{(x, u), (x, 2j - u), (y, v)\} \in D_3$.

(iiic) $\{(x, u), (y, v), (12, s)\} \notin \overline{\mathcal{A}}$ and $(x, y), (y, x) \notin \rho_k^1 \bigcup \rho_k^2$. There exists $z \in I_{12}$ such that $\{x, y, z\} \in \mathcal{J}_k$ by (A3). If $x < y < z$, there exist $j, w \in Z_n (j \neq s)$ such that $\{(x, u), (y, v), (12, j), (13, w)\} \in \overline{\mathcal{A}}$ then $P \subset \{(x, u), (y, v), (z, w')\} \in D_4$ where w' satisfies $\{(z, w'), (12, s), (13, w)\} \in \overline{\mathcal{A}}$. If $x < z < y$ (similar for $z < x < y$), there exists $w' \in Z_n$ such that $\{(y, v), (12, s), (13, w')\} \in \overline{\mathcal{A}}$, therefore there exist $j, w \in Z_n$ $(j \neq s)$ such that $\{(x, u), (z, w), (12, j), (13, w')\} \in \overline{\mathcal{A}}$, hence $P \subset \{(x, u), (y, v), (z, w)\} \in D_4$.

Step 2. Each $\mathcal{E}_s$ is an $STS(1 + 12n)$.

Direct calculation shows

$$|\mathcal{E}_s| = |E_1 \bigcup E_2| + |E_3| = \frac{2n(2n+1)}{6} \cdot 6 + 20n^2 = \frac{12n(12n+1)}{6}.$$

Therefore, we only need to show that every pair P from X is contained in some triple of $\mathcal{E}_s$. In fact,

(i) $P = \{\infty, (x, u)\}$. Since $\rho_{11}^2 = \phi$, there exists $y \in I_{12}$ such that (x, y) or $(y, x) \in \rho_{11}^1$ by (A2), thereby P is contained in E_1.

(ii) $P = \{(x, u), (x, v)\}, u \neq v$. As above there exists $y \in I_{12}$ such that (x, y) or $(y, x) \in \rho_{11}^1$, it follows that P is contained in $E_1 \bigcup E_2$.

(iii) $P = \{(x, u), (y, v)\}, x < y$. If (x, y) or $(y, x) \in \rho_{11}^1$ then P is contained in $E_1 \bigcup E_2$, or else there exists $z \in I_{12}$ such that $\{x, y, z\} \in \mathcal{J}_{11}$ and there exists $w \in Z_n$ such that $\{(x, u), (y, v), (z, w)\} \in \overline{\mathcal{A}}$ (when $y < z$) or $\{(x, u), (z, w), (y, \pi^{-s}(v)\} \in \overline{\mathcal{A}}$ (when $z < y$), therefore P is contained in E_3.

Step 3. $\{(X, \mathcal{D}_{ks}) : k \in I_{11}, s \in Z_n\} \bigcup \{(X, \mathcal{E}_s) : s \in Z_n \backslash \{0\}\}$ is an $LSTS(1 + 12n)$.

We only need to show that every triple T of distinct elements of X belongs to some $\mathcal{D}_{ks}$ or $\mathcal{E}_s$. All the posibilities are exhausted as follows:

(i) $T = \{\infty, (x, u), (x, v)\}, u \neq v$. By (A2) and (A5), there exists $y \in I_{12}$ such that (x, y) or $(y, x) \in \rho_{11}^1 = \rho_0^1 \bigcup \rho_1^1$. Here we only discuss the first case (the second case is similar). Let $\{\infty,(0,u),(0,v)\} \in \mathcal{B}_{is}$ by $LSTS(1+2n)$. If $i=0$, then T belongs to the part D_1 of $\mathcal{D}_{0s}$ or $\mathcal{D}_{1s}$; if $i=1$, then T belongs to the part E_1 of $\mathcal{E}_s$.

(ii) $T=\{(x,u),(x,v),(x,w)\}, u \neq v \neq w \neq u$. Similar to the case (i), T belongs either to the part D_2 of $\mathcal{D}_{0s}$ (or $\mathcal{D}_{1s}$), or to the part E_2 of $\mathcal{E}_s$, for certain $s \in Z_n$.

(iii) $T=\{\infty,(x,u),(y,v)\}, x \neq y$. If (x,y) or $(y,x) \in \rho_k^1$ for some $k \in I_{12}$, then $k=0,1,11$ by (A5). Therefore T belongs to D_1 of $\mathcal{D}_{0s}$ (or $\mathcal{D}_{1s}$) or E_1 of $\mathcal{E}_s$, for certain $s \in Z_n$. Or else, let $\{\infty,x,y\} \in C_k$ and $\{(x,u),(y,v),(12,s)\} \in \overline{\mathcal{A}}$, then T belongs to D_5 of $\mathcal{D}_{ks}$.

(iv) $T=\{(x,u),(x,v),(y,w)\}, x \neq y, u \neq v$. If (x,y) or $(y,x) \in \rho_k^1$ (of course $k=0,1,11$), then T belongs to D_2 of $\mathcal{D}_{0s}$ (or $\mathcal{D}_{1s}$) or E_2 of $\mathcal{E}_s$ for certain $s \in Z_n$. Or else, by (A6), there exists $k \in I_{11}$ such that $(x,y) \in \rho_k^2$. Let $\{(x,\frac{u+v}{2}),(y,w),(12,s)\} \in \overline{\mathcal{A}}$, then T belongs to the part D_3 of $\mathcal{D}_{ks}$.

(v) $T=\{(x,u),(y,v),(z,w)\}$, $x<y<z$. If $\{(x,u),(y,v),(z,w),(12,s)\} \in \overline{\mathcal{A}}$, let $\{x,y,z\} \in C_k$, then T belongs to the part D_6 of $\mathcal{D}_{ks}$. Or else $\{(x,u),(y,v),(z,w)\} \notin \overline{\mathcal{A}}$, let $\{x,y,z\} \in \mathcal{J}_k$. When $k=11$, let $\{(x,u),(y,v),(z,w')\} \in \overline{\mathcal{A}}$, $w' \neq w$, then we have $w=\pi^s(w')$ and $s \in Z_n \backslash \{0\}$, therefore T belongs to the part E_3 of $\mathcal{E}_s$. When $k \in I_{11}$, let $\{(x,u),(y,v),(12,j),(13,w')\} \in \overline{\mathcal{A}}$ and $\{(z,w),(12,s),(13,w')\} \in \overline{\mathcal{A}}$, then $j \neq s$ (otherwise it follows that $\{(x,u),(y,v),(z,w)\} \in \overline{\mathcal{A}}$, a contradiction), thus T belongs to the part D_4 of $\mathcal{D}_{ks}$. □

2.3.3 AD^*-design and $LSTS$

An AD^**−design* is a quintuple $AD^*\ (m)=AD^*[x]=(\beta_{01},\beta_{10},\beta_{11},\beta_{20},\beta_{21})$, where X is a $3m$-set, β_{01},β_{10} and β_{11} each consists of 3-subsets of X, and β_{20} and β_{21} each consists of 4-subsets of X such that the following conditions are satisfied:

(A^*1) These five sets of blocks (triples or quadruple) are pairwise disjoint;

(A^*2) Every 2-subset of X is contained in exactly one block of these five sets;

(A^*3) $\beta_{10} \bigcup \beta_{11}$ partitions X.

From the definition, it is easy to see that $|\beta_{10} \bigcup \beta_{11}| = m$ and $|\beta_{01}| + 2|\beta_{20} \bigcup \beta_{21}| = \frac{3}{2}m(m-1)$.

In what follows, F_q will denote the finite field of order q. In particular, $F_2 = \{0,1\} \subset F_4 = \{0,1,a,b\}$ and $F_4^* = F_4 \backslash \{0\}$, where $a^2 = b$, $b^2 = a$ and $1 + a + b = 0$. In this and next subsections the symbols a and b will be specially used for F_4.

For a subset B of X, $|B| \in \{2,3,4\}$, the notation ψ_B means a bijection from B onto F_2, F_4^* or F_4 when $|B| = 2, 3$ or 4, respectively. As an $AD^*[X]$ is given, a system of mappings ψ_B are given for each $B \in (\beta_{10} \bigcup \beta_{11} \bigcup \beta_{20} \bigcup \beta_{21})$. Further we define another system of mappings ψ'_B for $B \in (\beta_{10} \bigcup \beta_{11})$ by $\psi'_B(x) = 1$ (if $\psi_B(x) = a$) or $\psi'_B(x) = 0$ (if $\psi_B(x) = 1$ or b), where $\{1, a, b\} = F_4^*$ as mentioned above.

Define $LAD_1(m) = LAD_1[X] = \{ AD_j^*(m)\colon\ j \in I_{3m-1}\}$, where X is a $3m$-set and $AD_j^*(m) = (\beta_{01}^j, \beta_{10}^j, \beta_{11}^j, \beta_{20}^j, \beta_{21}^j)$ such that, besides $(A^*1) - (A^*3)$ for each AD_j^*, the following additional conditions are satisfied:

(A^*4) For any triple T of X, there exist j, i, y and B such that $T \subseteq B \in \beta_{iy}^j$;

(A^*5) $\bigcup_{j \in I_{3m-1}} \beta_{i0}^j = \bigcup_{j \in I_{3m-1}} \beta_{i1}^j$, for $i = 1, 2$;

(A^*6) For any 2-subset P of X, there exist a unique j and a unique B such that $P \subset B \in \beta_{10}^j$.

It is easy to see that $|\bigcup_j \beta_{10}^j| = |\bigcup_j \beta_{11}^j| = \frac{m}{2}(3m-1)$.

Define $LAD_2(m) = LAD_2[X] = \{\gamma, AD_j^*(m)\colon\ j \in I_{3m-2}\}$, where X is a $3m$-set (m even), γ is a partition of X into 2-subsets and $AD_j^*(m) = (\beta_{01}^j, \beta_{10}^j, \beta_{11}^j, \beta_{20}^j, \beta_{21}^j)$ such that, besides $(A^*1) - (A^*3)$ for each AD_j^*, the following additional conditions are satisfied:

(A^*7) For each $j \in I_{3m-2}$, $\beta_{11}^j = \phi$;

(A^*8) For any triple T of X, there exist j, i, y and B such that $T \subseteq B \in \beta_{iy}^j$;

(A^*9) $\bigcup_{j \in I_{3m-2}} \beta_{20}^j = \bigcup_{j \in I_{3m-2}} \beta_{21}^j$;

(A^*10) For any 2-subset P of X and $P \notin \gamma$, there exist two blocks B and B' in $\bigcup_{j \in I_{3m-2}} \beta_{10}^j$ such that $B \bigcap B' = P$;

(A^*11) The system of mappings $\{\psi_B;\, B \in \gamma \bigcup_{j\in I_{3m-2}} (\beta^j_{10} \bigcup \beta^j_{20} \bigcup \beta^j_{21})\}$ satisfies that if both A and B belong to $\bigcup_{j\in I_{3m-2}} \beta^j_{10}$ and $A \bigcap B = \{x, y\}$ $(x \neq y)$, then $\psi_A(x)\psi_B(y) \neq \psi_A(y)\psi_B(x)$;

(A^*12) There is another system of mappings $\{\phi_{uB} : u \in F_4$ and $B \in \bigcup_{j\in I_{3m-2}} \beta^j_{10}\}$, where $\phi_{uB} : B \to F_2$, such that

(i) if $A, B \in \bigcup_{j\in I_{3m-2}} \beta^j_{10}$ and $A \bigcap B = \{x, y\}$, $x \neq y$, then $\phi_{uA}(x) + \phi_{uA}(y) + \phi_{uB}(x) + \phi_{uB}(y) = 1$ for any $u \in F_4$;

(ii) for any $u \neq v \in F_4$ and $B \in \bigcup_{j\in I_{3m-2}} \beta^j_{10}$ there exists no y satisfying the simultaneous equations $\{\phi_{uB}(x) + \phi_{vB}(x) = y$: $x \in B\}$.

It is easy to see that $|\beta^j_{10}| = m$ for each $j \in I_{3m-2}$, and $P \not\subset B$ for any pair $P \in \gamma$ and any $B \in \bigcup_j \beta^j_{10}$.

Let X be an n-set. We shall need to decompose $n(2n-1)$ unordered pairs from the set $F_2 \times X$ into $2n$ system $\gamma_{yx}[X]$ $(y \in F_2, x \in X)$ such that each system $\gamma_{0x}[X]$ contains $n-1$ mutually disjoint pairs without $(0, x)$ and $(1, x)$, and each $\gamma_{1x}[X]$ contains n mutually disjoint pairs. For any positive integer n such a decomposition always exists. H. Hanani has given a construction for $\gamma_{yx}[X]$ [Ha].

The AD^*–design and its large sets LAD_1 and LAD_2 had been introduced by Lu Jiaxi [Lu 4], and he gave the following examples, which, as we shall see, will play important role to construct some $LSTS$.

An $LAD_1(1)$: Define $LAD_1(1) = LAD_1[I_3] = \{AD^*_0(1), AD^*_1(1)\}$, where $AD^*_i(1) = (\beta^i_{01}, \beta^i_{10}, \beta^i_{11}, \beta^i_{20}, \beta^i_{21})$ and $\beta^0_{10} = \beta^1_{11} = \{\{0, 1, 2\}\}$, other eight β^i_{jk} are all empty. It is easy to verify that the construction satisfies (A^*_1)—(A^*_6).

An $LAD_1(5)$: First, construct a 3-(16, 4, 1) design, which can be decomposed into seven $B[4, 1; 16]$, denoted by $(I_2 \times F_8, \mathcal{B}_x)$, $x \in F^*_8$. Each $\mathcal{B}_x$ consists of the following blocks (in which g is a primitive element of

F_8, $g^3 = 1+g$, i runs over I_2, and y runs over F_8):

$$\begin{aligned}
&\{(i,0),(i,xg),(i,xg^2),(i,xg^4)\};\\
&\{(i,x),(i,xg^3),(i,xg^5),(i,xg^6)\};\\
&\{(0,xy),(0,x+xy),(1,xy),(1,x+xy)\};\\
&\{(0,xy),(0,xg^3+xy),(1,xg^2+xy),(1,xg^5+xy)\};\\
&\{(0,xy),(0,xg^5+xy),(1,xg+xy),(1,xg^6+xy)\};\\
&\{(0,xy),(0,xg^6+xy),(1,xg^4+xy),(1,xg^3+xy)\}.
\end{aligned}$$

The first and second family each contains two quadruples, and other four families each contains four different quadruples.

Define $LAD_1(5) = LAD_1[(I_2 \times F_8)\backslash\{(0,0)\}] = \{AD_j^*(5) : j \in Z_{14}\}$, where $AD_j^*(5) = (\beta_{01}^i, \beta_{10}^i, \beta_{11}^i, \beta_{20}^i, \beta_{21}^i)$. For $0 \le j \le 6$, define

$\beta_{10}^j = \{\{(x,u),(y,v),(z,w)\} : \{(0,0),(x,u),(y,v),(z,w)\} \in \mathcal{B}_{g^j}\}$,

$\beta_{20}^j = \{B : B \in \mathcal{B}_{g^j}, (0,0) \notin B\}$,

$\beta_{01}^j = \beta_{11}^j = \beta_{21}^j = \phi$;

and $\beta_{i0}^{j+7} = \beta_{i1}^j$, $\beta_{i1}^{j+7} = \beta_{i0}^j$, $0 \le j \le 6$ and $0 \le i \le 2$ (here $\beta_{00}^j = \phi, 0 \le j \le 6$).

An $LAD_2(4)$: Let $LAD_2(4) = LAD_2[I_6 \times F_2] = \{\gamma, AD_j^*(4) : j \in I_{10}\}$, where $\gamma = \{\{(i,0),(i,1)\} : i \in I_6\}$ and $AD_j^*(4) = (\beta_{01}^j, \beta_{10}^j, \beta_{11}^j, \beta_{20}^j, \beta_{21}^j)$, $j \in I_{10}$. All $\beta_{10}^j (j \in I_{10})$ are listed in the following Table 2.3. Each β_{10}^j contains four triples, and the mappings $\phi_{uB}(u \in F_4, B \in \beta_{10}^j)$ satisfying (A^*12) are shown under both β_{10}^{2j} and $\beta_{10}^{2j+1} (j \in I_5)$. In the last row of the table the mappings ψ_B $(B \in \beta_{10}^j, j \in I_{10})$ satisfying (A^*11) are shown. And, $\beta_{01}^j = \{\{(x_0,y_0),(x_1,y_1),(x_2,y_2)\}$: $\{(x_0,z_0),(x_1,z_1),(x_2,z_2)\} \in \beta_{10}^j$ and $y_0+y_1+y_2 = z_0+z_1+z_2\}$, $j \in I_{10}$. It is easy to see that $|\beta_{10}^j| = 4$ and $|\beta_{01}^j| = 12$ for $j \in I_{10}$. As well as,

$\beta_{11}^j = \phi$, for $j \in I_{10}$;

$\beta_{20}^{1+2i} = \beta_{21}^{2i} = \phi$, for $i \in I_5$;

$\beta_{20}^0 = \beta_{21}^1 : \{(0,0),(0,1),(1,0),(1,1)\}, \{(2,0),(2,1),(3,0),(3,1)\}, \{(4,0),(4,1),(5,0),(5,1)\}$;

$\beta_{20}^2 = \beta_{21}^3 : \{(0,0),(0,1),(2,0),(2,1)\}, \{(1,0),(1,1),(4,0),(4,1)\}, \{(3,0),(3,1),(5,0),(5,1)\}$;

$\beta_{20}^4 = \beta_{21}^5 : \{(0,0),(0,1),(3,0),(3,1)\}, \{(1,0),(1,1),(5,0),(5,1)\}, \{(2,0),(2,1),(4,0),(4,1)\}$;

$\beta_{20}^{6} = \beta_{21}^{7} : \{(0,0),(0,1),(4,0),(4,1)\},\{(1,0),(1,1),(3,0),(3,1)\},$
$\{(2,0),(2,1),(5,0),(5,1)\};$

$\beta_{20}^{8} = \beta_{21}^{9} : \{(0,0),(0,1),(5,0),(5,1)\},\{(1,0),(1,1),(2,0),(2,1)\},$
$\{(3,0),(3,1),(4,0),(4,1)\}.$

Table 2.3 β_{10}^{j} **in** $LAD_2(4)$

β_{10}^{0}	(0,0) (2,0) (4,0)	(0,1) (3,0) (5,1)	(1,0) (3,1) (4,1)	(1,1) (2,1) (5,0)
β_{10}^{1}	(1,1) (4,1) (2,0)	(1,0) (5,0) (3,0)	(0,0) (5,1) (2,1)	(0,1) (4,0) (3,1)
ϕ_{0B}	0 0 0	0 0 1	1 0 0	1 0 0
ϕ_{1B}	1 0 0	0 1 0	0 0 1	0 1 0
ϕ_{aB}	0 0 1	1 0 0	0 1 0	0 0 0
ϕ_{bB}	0 1 0	0 0 0	0 0 0	0 0 1
β_{10}^{2}	(0,1) (1,1) (3,0)	(0,0) (4,0) (5,0)	(1,0) (5,1) (2,0)	(2,1) (4,1) (3,1)
β_{10}^{3}	(1,0) (0,1) (5,0)	(1,1) (2,0) (3,1)	(0,0) (3,0) (4,1)	(4,0) (2,1) (5,1)
ϕ_{0B}	0 0 1	0 1 0	0 1 0	0 1 0
ϕ_{1B}	0 1 0	0 0 1	0 0 0	0 0 0
ϕ_{aB}	0 0 0	0 0 0	1 0 0	1 0 0
ϕ_{bB}	1 0 0	1 0 0	0 0 1	0 0 1
β_{10}^{4}	(0,0) (1,0) (2,0)	(0,1) (5,1) (4,1)	(3,0) (1,1) (4,0)	(2,1) (3,1) (5,0)
β_{10}^{5}	(1,1) (0,0) (4,1)	(1,0) (3,0) (2,1)	(5,0) (0,1) (2,0)	(4,0) (5,1) (3,1)
ϕ_{0B}	0 0 1	0 0 1	0 0 1	1 0 0
ϕ_{1B}	0 1 0	1 0 0	0 0 0	0 1 0
ϕ_{aB}	1 0 0	0 0 0	0 1 0	0 0 1
ϕ_{bB}	0 0 0	0 1 0	1 0 0	0 0 0
β_{10}^{6}	(0,1) (2,1) (1,1)	(0,0) (5,0) (3,1)	(1,0) (4,1) (5,1)	(2,0) (3,0) (4,0)
β_{10}^{9}	(1,0) (4,0) (0,1)	(1,1) (3,1) (5,1)	(0,0) (2,1) (3,0)	(4,1) (5,0) (2,0)
ϕ_{0B}	1 0 0	0 1 0	0 0 0	1 0 0
ϕ_{1B}	0 0 0	0 0 0	0 0 1	0 0 1
ϕ_{aB}	0 0 1	1 0 0	0 1 0	0 1 0
ϕ_{bB}	0 1 0	0 0 1	1 0 0	0 0 0
β_{10}^{8}	(0,0) (3,1) (1,0)	(2,1) (0,1) (4,1)	(1,1) (5,0) (4,0)	(2,0) (5,1) (3,0)
β_{10}^{7}	(1,1) (5,1) (0,0)	(4,0) (1,0) (2,1)	(0,1) (3,1) (2,0)	(4,1) (3,0) (5,0)
ϕ_{0B}	1 0 0	0 0 0	0 0 0	0 0 0
ϕ_{1B}	0 1 0	1 0 0	1 0 0	1 0 0
ϕ_{aB}	0 0 0	0 1 0	0 1 0	0 1 0
ϕ_{bB}	0 0 1	0 0 1	0 0 1	0 0 1

Theorem 2.3.3. For positive integers $m, n (n \neq 1, 3)$, if there exist both an $LAD_1(m)$ and an $LSTS(1+4n)$, then there exists an $LSTS(1+12mn)$.

Construction. Let X be a $3m$-set, and

$LAD_1(m) = LAD_1[X] = \{AD_j^*(m) : j \in I_{3m-1}\}$ with mapping systems $\{\psi_B\}$ and $\{\psi'_B\}$, where $AD_j^*(m) = (\beta_{01}^j, \beta_{10}^j, \beta_{11}^j, \beta_{20}^j, \beta_{21}^j)$;

$LSTS(1+4n) = \{(\{\infty\} \bigcup (F_2 \times F_2 \times Z_n), \mathcal{B}_{yzw})\colon (y,z,w) \in (F_2 \times F_2 \times Z_n) \backslash \{(0,0,0)\}\}$;

$\gamma_{y,(z,w)}[F_2 \times Z_n]$, where $y \in F_2$ and $(z,w) \in F_2 \times Z_n$.

Since $n \neq 1, 3$, thus there exists a $TD(4, 2n) = (F_4 \times F_2 \times Z_n, \mathcal{G}, \mathcal{A})$, where $\mathcal{G} = \{\{u\} \times F_2 \times Z_n : u \in F_4\}$. Below, the expression $B \in \overline{\mathcal{A}}$ will denote the fact there is a block A such that $B \subset A \in \mathcal{A}$.

First, construct $4n(3m-1)$ triple system Ω_{yzw}^j on the set $S = \{\infty\} \bigcup (X \times F_2 \times F_2 \times Z_n)$, where $j \in I_{3m-1}$ and $(y,z,w) \in F_2 \times F_2 \times Z_n$. Each Ω_{yzw}^j consists of the following eight parts.

O_1: $\{\infty, (x_0, y_0, z_0, w_0), (x_1, y_1, z_1, w)\}$ provided there exists x_2, y', z_2 and w_2 such that $B = \{x_0, x_1, x_2\} \in \beta_{1y'}^j$, $\psi_B(x_1) = a\psi_B(x_0)$, $\{(\psi_B(x_0), z_0 + y'\psi'_B(x_2), w_0), (\psi_B(x_1), z_1 + y'\psi'_B(x_0), w_1), (\psi_B(x_2), z_2 + y'\psi'_B(x_1), w_2), (0, z, w)\} \in \mathcal{A}$, $y_0 = y + y' + z_1\psi'_B(x_0) + y'\psi'_B(x_2)$ and $y_1 = 1 + y + y' + z_2\psi'_B(x_1) + y'\psi'_B(x_0)$.

$O_2 - O_8$: $\{(x_0, y_0, z_0, w_0), (x_1, y_1, z_1, w_1), (x_2, y_2, z_2, w_2)\}$, where the conditions for each part are as follows:

O_2: $\{x_0, x_1, x_2\} \in \beta_{01}^j$, $y_0 + y_1 + y_2 = y$, $z_0 + z_1 + z_2 = z$ and $w_0 + w_1 + w_2 = w$;

O_3: There exist B and y' such that $\{x_0, x_1, x_2\} \subset B \in \beta_{2y'}^j$, $\psi_B(x_i) = a^i (i \in I_3)$, $y_0 + y_1 = y$, $y_0 + y_2 = y'$, $z_0 + z_1 + z_2 = z$ and $w_0 + w_1 + w_2 = w$;

O_4: There exist B and y' such that $\{x_0, x_1, x_2\} \subset B \in \beta_{2y'}^j$, $\psi_B(x_i) = a^i (i \in I_2)$, $\psi_B(x_2) = 0$, $y_0 + y_1 = y + 1$, $y_0 + y_2 = y + y'$, $z_0 + z_1 + z_2 = z$ and $w_0 + w_1 + w_2 = w$;

O_5: There exist B and y' such that $\{x_0, x_1, x_2\} \subset B \in \beta_{2y'}^j$, $\psi_B(x_i) = a^{i+2} (i \in I_2)$, $\psi_B(x_2) = 0$, $y_0 + y_1 = y' + 1$, $y_1 + y_2 = y + y' + 1$, $z_0 + z_1 + z_2 = z$ and $w_0 + w_1 + w_2 = w$;

O_6: There exist B and y' such that $\{x_0, x_1, x_2\} \subset B \in \beta_{2y'}^j$, $\psi_B(x_i) = a^{i+1} (i \in I_2)$, $\psi_B(x_2) = 0$, $y_0 + y_1 = y + y' + 1$, $y_0 + y_2 = y'$, $z_0 + z_1 + z_2 = z$ and $w_0 + w_1 + w_2 = w$;

O_7: There exist $x_3, y', y'', z', z_3, w'$ and w_3 such that $x_1 = x_2$, $B = \{x_0, x_1, x_3\} \in \beta_{1y''}^j$, $\psi_B(x_0) = (b + y'')\psi_B(x_1)$, $\{(y_1, z_1, w_1), (y_2, z_2, w_2)\} \in$

$\gamma_{y',(z',w')}[F_2\times Z_n]$, $\{(\psi_B(x_0), z_0+y''\psi'_B(x_1), w_0), (\psi_B(x_1), z'+y''\psi'_B(x_3), w'), (\psi_B(x_3), z_3 + y''\psi'_B(x_0), w_3), (0, z, w)\} \in \mathcal{A}$ and $y_0 + y + y' + z'\psi'_B(x_0) + y''\psi'_B(x_1) + y''(z' + z_3)\psi'_B(x_0) = 0$;

O_8: There exists y' such that $B = \{x_0, x_1, x_2\} \in \beta^j_{1y'}$, $\psi_B(x_2) = a\psi_B(x_1) = b\psi_B(x_0)$, $\{(\psi_B(x_0), z_0 + y'\psi'_B(x_2), w_0), (\psi_B(x_1), z_1 + y'\psi'_B(x_0), w_1), (\psi_B(x_2), z_2 + y'\psi'_B(x_1), w_2), (0, z, w)\} \in \mathcal{A}$, $y_0 = y + z_1\psi'_B(x_0) + y'\psi'_B(x_2)$, $y_1 = y + z_2\psi'_B(x_1) + y'\psi'_B(x_0)$ and $y_2 = y + z_0\psi'_B(x_2) + y'\psi'_B(x_1)$.

Furthermore, construct $4n-1$ triple system $\mathcal{T}_{yzw}$ on the set S, where $(y,z,w) \in (F_2 \times F_2 \times Z_n)\backslash\{(0,0,0)\}$. Each $\mathcal{T}_{yzw}$ consists of the following three parts.

T_1: $\{\infty, (x, y_0, z_0, w_0), (x, y_1, z_1, w_1)\}$ provided $x \in X$ and $\{\infty, (y_0, z_0, w_0), (y_1, z_1, w_1)\} \in \mathcal{B}_{yzw}$.

T_2: $\{(x, y_0, z_0, w_0), (x, y_1, z_1, w_1), (x, y_2, z_2, w_2)\}$ provided $x \in X$ and $\{(y_0, z_0, w_0), (y_1, z_1, w_1), (y_2, z_2, w_2)\} \in \mathcal{B}_{yzw}$.

T_3: $\{(x_0, y_0, z_0, w_0), (x_1, y_1, z_1, w_1), (x_2, y_2, z_2, w_2)\}$ provided $B = \{x_0, x_1, x_2\} \in \bigcup_{j\in I_{3m-1}} \beta^j_{10}$, $\psi_B(x_i) = a^i$ $(i \in I_3)$, $\{(1, z_0, w_0), (a, z_1, w_1), (b, y + y_1 + y_2, w + w_2)\} \in \overline{\mathcal{A}}$ and $y_0 + y_1 + z_2 + z = 0$.

Then, $\{(S, \Omega^j_{yzw})\colon j \in I_{3m-1}, (y,z,w) \in F_2\times F_2\times Z_n\} \bigcup \{(S, \mathcal{T}_{yzw})\colon (y, z, w) \in (F_2 \times F_2 \times Z_n)\backslash\{(0,0,0)\}\}$ is an $LSTS(1+12mn)$.

Proof. It is not hard to calculate that

$|\Omega^j_{yzw}| = |O_1| + |O_2| + \sum_{i=3}^{6} |O_i| + |O_7| + |O_8| = 3|\beta^j_{10} \bigcup \beta^j_{11}||F_2||Z_n| + |\beta^j_{01}|2^2 \cdot 2^2 \cdot n^2 + 4(|\beta^j_{20} \bigcup \beta^j_{21}||F_2|2^2 \cdot n^2) + 6|\beta^j_{10} \bigcup \beta^j_{11}|n(4n-1) + 2n|\beta^j_{10} \bigcup \beta^j_{11}| = 6mn + 16n^2\frac{3m}{2}(m-1) + 6mn(4n-1) + 2mn = \frac{12mn(12mn+1)}{6}$,

$|\mathcal{T}_{yzw}| = |T_1| + |T_2| + |T_3| = 6mn + 3m(\frac{4n(4n+1)}{6} - 2n) + 8mn^2(3m-1) = \frac{12mn(12mn+1)}{6}$.

Further work is tedious, which is omitted here (refer to [Lu 4]). □

Corollary 2.3.4. For positive integer n, if there exists an $LSTS(1+4n)$ then there exists an $LSTS(1+12n)$.

Proof. This following from Theorem 2.3.3 and an $LAD_1(1)$ if $n \neq 1, 3$. But, for $n = 1$, there exists no $LSTS(1+4)$. And for $n = 3$ there exists an $LSTS(37)$ as follows (refer to [Lu 1]):

First, define 7 ordered triple systems (C_{0t}, ϕ_{0t}) on the set $\{\infty_1, \infty_2\} \bigcup$

Z_5, where each $C_{0t}(t \in Z_7)$ contains $(a, b, 0)$ and other six blocks, which are listed in the first column of Table 2.4 corresponding six nonempty positions in column C_{0t}; and each ϕ_{0t} is a mapping $B \to Z_7$ for each $B \in C_{0t} \backslash \{(a, b, 0)\}$, which, as an entry, appears in column C_{0t}. For example, the system C_{00} contains triples $(a, 3, 4)2, (a, 2, 1)0, (b, 1, 3)1, (b, 4, 2)2$, $(0, 1, 4)0$ and $(0, 3, 2)0$, where the attaching number after each triple B is the mapping $\phi_{00}(B)$.

Table 2.4 (C_{0t}, ϕ_{0t})

B	C_{00}	C_{01}	C_{02}	C_{03}	C_{04}	C_{05}	C_{06}
a 1 3			5		4		
a 2 4			2				
a 4 1						0	
a 3 1							6
a 4 2					4		1
a 2 3						5	
a 3 4	2			4			
a 2 1	0	1		4			
a 4 3		6					
b 1 3	1			3			
b 4 1			6				
b 1 4		5					0
b 4 2	2			3			
b 1 2						6	
b 2 3							0
b 3 4					3	1	
b 2 1					3		
b 3 2		2	5				
0 1 4	0			3	4		
0 3 4			2				1
0 2 1			5				6
0 3 2	0			3	4		
0 1 3		1				2	
0 4 2		6				5	

Furthermore, define $C_{st} = \{(x+s, y+s, z+s)\colon (x, y, z) \in C_{0t}\}$ and $\phi_{st}(x+s, y+s, z+s) = \phi_{0t}(x, y, z), s \in Z_5, t \in Z_7$, where $\infty_1 + s = \infty_1$, and $\infty_2 + s = \infty_2$ for any $s \in Z_5$, Then it is easy to see that

(1) Each $\overline{C}_{st} = \{\{x, y, z\} : (x, y, z) \in C_{st}\}$ is an $STS(7)$ on the set $\{\infty_1, \infty_2\} \bigcup Z_5$, for any $(s, t) \in Z_5 \times Z_7$;

(2) $\{x \in Z_5\colon\ \exists y \in Z_5 \bigcup \{\infty_1, \infty_2\}$ such that (a, x, y) or $(b, x, y) \in C_{st}\} = Z_5$ for any $(s, t) \in Z_5 \times Z_7$;

(3) $\{(x, y) \in Z_5 \times Z_5 : \exists s \in Z_5$ such that (a, x, y) or $(b, x, y) \in C_{s0}\} = \{(x, y) \in Z_5 \times Z_5\colon\ x \neq y\}$.

Let $LSTS(9) = \{(\{\infty_1, \infty_2\} \bigcup Z_7, \mathcal{A}_t) : t \in Z_7\}$. Construct $LSTS(37) = \{(\{\infty_1, \infty_2\} \bigcup (Z_5 \times Z_7),\ \mathcal{B}_{st})\colon\ s \in Z_5, t \in Z_7\}$, where each $\mathcal{B}_{st}$ constists of the following triples :

part 1. $\{(i, x), (i, y), (i, z)\}$ with $i \in Z_5$ and $\{x, y, z\} \in \mathcal{A}_t$, whenever ∞_1 or ∞_2 appears for x, y or z , omit the index i before it .

part 2. $\{(i, x), (i, y), (j, z)\}$ with $(u, i, j) \in C_{st}$, $u = \infty_1$ or ∞_2, $x \neq y$ and $x + y = 2(z + \phi(u, i, j))$.

part 3. $\{u, (i, x), (j, y)\}$ with $(u, i, j) \in C_{st}$, $u = \infty_1$ or ∞_2 and $x - y = \phi(u, i, j)$.

part 4. $\{(i, x), (j, y), (k, z)\}$ with $(i, j, k) \in C_{st}$ and $x+y+z = \phi(i, j, k)$.

□

Corollary 2.3.5. For positive integer n, if there exists an $LSTS(1+4n)$, then there exists an $LSTS(1 + 60n)$.

Proof. This follows from Theorem 2.3.3 and an $LAD_1(5)$ if $n \neq 1, 3$. But, for $n = 1$, there exists no $LSTS(1 + 4)$. And, for $n = 3$, there exists an $LSTS(181)$: the existence of $LSTS(9)$ and $LD(7)$ implies the existence of $LSTS(21)$ by Theorem 2.3.1, furthermore $LSTS(21) \to LSTS(61) \to LSTS(181)$ by Corollary 2.3.4 (taking $n = 5$ and 15). □

Theorem 2.3.6. For positive integer $n (\neq 2, 6)$ and even m, if there exist both an $LAD_2(m)$ and an $LSTS(1 + 4n)$, then there exists an $LSTS(1 + 6mn)$.

Construction. Let X be a $3m$-set, and

$LAD_2(m) = LAD_2[X] = \{\gamma, AD_j^*(m) : j \in I_{3m-2}\}$, with mapping systems $\{\psi_B\}$ and $\{\phi_{uB}\}$, where $AD_j^*(m) = (\beta_{01}^j, \beta_{10}^j, \beta_{20}^j, \beta_{21}^j)$;

$LSTS(1 + 4n) = \{(\{\infty\} \bigcup (F_2 \times F_2 \times Z_n),\ \mathcal{B}_{uw})\colon (u, w) \in (F_4 \times Z_n) \backslash \{(0, 0)\}\}$;

$\{\gamma_{yw}[Z_n]\colon\ y \in F_2, w \in Z_n\}$.

Since $n \neq 2, 6$, thus there exists a $TD(4, n) = (F_4 \times Z_n, \mathcal{G}, \mathcal{A})$, where $\mathcal{G} = \{\{u\} \times Z_n\colon\ u \in F_4\}$. Below, the expression $B \in \overline{\mathcal{A}}$ will denote the fact that there is a block A such that $B \subset A \in \overline{\mathcal{A}}$.

First, construct $2n(3m-2)$ triple systems $\mathcal{K}^j_{yw}$ on the set $S = \{\infty\} \bigcup (X \times F_2 \times Z_n)$, where $j \in I_{3m-2}$ and $(y,w) \in F_2 \times Z_n$. Each $\mathcal{K}^j_{yw}$ consists of the following eight parts.

K_1:$\{\infty, (x_0, y_0, w_0), (x_1, y_1, w_1)\}$ provided there is B such that $\{x_0, x_1\} \subset B \in \beta^j_{10}$, $\psi_B(x_1) = a\psi_B(x_0)$, $y_0 = \phi_{0B}(x_0) + y, y_1 = \phi_{0B}(x_1) + y + 1$ and $\{(\psi_B(x_0), w_0), (\psi_B(x_1), w_1), (0, w)\} \in \overline{\mathcal{A}}$.

$K_2 - K_8 : \{(x_0, y_0, w_0), (x_1, y_1, w_1), (x_2, y_2, w_2)\}$, where the conditions for each part are as follows:

$K_2 : \{x_0, x_1, x_2\} \in \beta^j_{01}$, $y_0 + y_1 + y_2 = y$ and $w_0 + w_1 + w_2 = w$.

K_3 : There exist B and y' such that $\{x_0, x_1, x_2\} \subset B \in \beta^j_{2y'}$, $\psi_B(x_i) = a^i$ $(i \in I_3)$, $y_0 + y_1 = y, y_0 + y_2 = y'$ and $w_0 + w_1 + w_2 = w$.

K_4 : There exist B and y' such that $\{x_0, x_1, x_2\} \subset B \in \beta^j_{2y'}$, $\psi_B(x_i) = a^i$ $(i \in I_2)$, $\psi_B(x_2) = 0, y_0 + y_1 = y + 1, y_0 + y_2 = y + y'$ and $w_0 + w_1 + w_2 = w$.

K_5 : There exist B and y' such that $\{x_0, x_1, x_2\} \subset B \in \beta^j_{2y'}$, $\psi_B(x_i) = a^{i+2}$ $(i \in I_2)$, $\psi_B(x_2) = 0, y_0 + y_1 = y' + 1, y_1 + y_2 = y + y' + 1$ and $w_0 + w_1 + w_2 = w$.

K_6 : There exist B and y' such that $\{x_0, x_1, x_2\} \subset B \in \beta^j_{2y'}$, $\psi_B(x_i) = a^{i+1}$ $(i \in I_2)$, $\psi_B(x_2) = 0, y_0 + y_1 = y + y' + 1, y_0 + y_2 = y'$ and $w_0 + w_1 + w_2 = w$.

K_7 : There exist B and y', w' such that $\{x_0, x_1\} \subset B \in \beta^j_{10}$, $x_1 = x_2$, $\psi_B(x_1) = a\psi_B(x_0)$, $\{(y_1, w_1), (y_2, w_2)\} \in \gamma_{y'w'}[Z_n]$, $\{(\psi_B(x_0), w_0), (\psi_B(x_1), w'), (0, w)\} \in \overline{\mathcal{A}}$ and $y_0 + y + y' + \phi_{0B}(x_0) = 0$.

K_8 : $B = \{x_0, x_1, x_2\} \in \beta^j_{10}$, $y_i + y = \phi_{0B}(x_i)$ $(i \in I_3)$ and $\{(\psi_B(x_0), w_0), (\psi_B(x_1), w_1), (\psi_B(x_2), w_2), (0, w)\} \in \mathcal{A}$.

Furthermore, construct $4n - 1$ triple systems $\mathcal{F}_{uw}$ on the set S, where $(u, w) \in (F_4 \times Z_n \backslash \{(0,0)\}$. Each $\mathcal{F}_{uw}$ consists of the following three parts.

F_1 : $\{\infty, (x_0, y_0, w_0), (x_1, y_1, w_1)\}$ provided there is B such that $\{x_0, x_1\} \subseteq B \in \gamma$ and $\{\infty, (\psi_B(x_0), y_0, w_0), (\psi_B(x_1), y_1, w_1)\} \in \mathcal{B}_{uw}$.

F_2 : $\{(x_0, y_0, w_0), (x_1, y_1, w_1), (x_1, y_2, w_2)\}$ provided there is B such that $\{x_0, x_1\} \subseteq B \in \gamma$ and $\{(\psi_B(x_0), y_0, w_0), (\psi_B(x_1), y_1, w_1), (\psi_B(x_1), y_2, w_2)\} \in \mathcal{B}_{uw}$.

$F_3 : \{(x_0, y_0, w_0), (x_1, y_1, w_1), (x_2, y_2, w_2)\}$ provided $B = \{x_0, x_1, x_2\} \in \bigcup_{j \in I_{3m-2}} \beta_{10}^j$, $\psi_B(x_i) = a^i$ $(i \in I_3), y_i = \phi_{uB}(x_i) + y$ $(i \in I_3, y \in F_2)$ and $\{(1, w_0), (a, w_1), (b, w + w_2)\} \in \overline{\mathcal{A}}$. (Note that in the expression $\{x_0, x_1\} \subset B \in \beta_{10}^j$ (or $\{x_0, x_1\} \subseteq B \in \gamma$), $x_0 = x_1$ may be permited.)

Then, $\{(S, \mathcal{K}_{yw}^j)$: $j \in I_{3m-2}$, $(y, w) \in F_2 \times Z_n\} \bigcup \{(S, \mathcal{F}_{uw})$: $(u, w) \in (F_4 \times Z_n) \backslash \{(0, 0)\}\}$ will be an $LSTS(1 + 6mn)$.

Proof. It is not hard to calculate that

$$\begin{aligned}|\mathcal{K}_{yw}^j| &= |K_1| + |K_2| + \sum_{i=3}^{6} |K_i| + |K_7| + |K_8| \\ &= 3n|\beta_{10}^j| + 2(4n^2|\beta_{01}^j|) + 4(2n^2|\beta_{20}^j \cup \beta_{21}^j|) + 3n(2n-1)|\beta_{10}^j| + n|\beta_{10}^j| \\ &= 3mn + 4n^2 \cdot \tfrac{3}{2}m(m-1) + 3mn(2n-1) + mn \\ &= \tfrac{6mn(6mn+1)}{6},\end{aligned}$$

$$|\mathcal{F}_{uw}| = |F_1 \cup F_2| + |F_3| = mn(4n+1) + 2mn^2(3m-2) = \tfrac{6mn(6mn+1)}{6}.$$

Further work is tedious (some step is similar to Theorem 2.3.3), which is omitted here. (refer to [Lu 4]). □

Corollary 2.3.7. For positive interger n, if there exists an $LSTS(1+4n)$ then there exists an $LSTS(1 + 24n)$.

Proof. This follows from Theorem 2.3.6 and $LAD_2(4)$ if $n \neq 2, 6$. For $n = 2$, there exists an $LSTS(49)$ by [Den 2]. And, for $n = 6$, $LSTS(49) \to LSTS(145)$ by Corollary 2.3.4 (taking $n = 12$). □

2.3.4 AD^{**}-design and $LSTS$

An AD^{**}-design is a quintuple $AD^{**}(m) = AD^{**}[X] = (\beta_{10}, \beta_{11}, \beta_{a0}, \beta_{a1}, \beta_{b1})$, where X is a $3m$-set and each $\beta_{vy}, (v, y) \in (F_4^* \times F_2) \backslash \{(b, 0)\}$, consists of triples of X such that the following conditions are satisfied:

($A^{**}1$) These five sets of triples are pairwise disjoint;

($A^{**}2$) Every 2-subset of X is contained in exactly one triple of these five sets;

($A^{**}3$) $\beta_{10} \cup \beta_{a0} \cup \beta_{a1} \cup \beta_{b1}$ partitions X.

From these conditions we can get $|\beta_{10} \cup \beta_{a0} \cup \beta_{a1} \cup \beta_{b1}| = m$ and $|\beta_{11}| = \frac{3m}{2}(m-1)$. When an $AD^{**}[X]$ is given, as in subsection 2.3.3, we set up a system of bijections $\psi_B : B \to F_4^*$ for each $B \in (\beta_{10} \cup \beta_{a0} \cup \beta_{a1} \cup \beta_{b1})$.

Define $LAD_3(m) = LAD_3[X] = \{AD_j^{**}(m)\colon j \in I_{3m-2}\}$, where X is a $3m$-set and $AD_j^{**}(m) = (\beta_{10}^j, \beta_{11}^j, \beta_{a0}^j, \beta_{a1}^j, \beta_{b1}^j)$ such that, besides $(A^{**}1) - (A^{**}3)$ for each AD_j^{**}, the following additional conditions are satisfied:

$(A^{**}4)$ For any triple T of the set X there exists unique (j, v, y) such that $T \in \beta_{vy}^j$;

$(A^{**}5)$ $\bigcup_{j \in I_{3m-2}} \beta_{10}^j$ partitions X;

$(A^{**}6)$ For any 2-subset P of X, either P is contained in some triple of $\bigcup_j \beta_{10}^j$ or P is the intersection of two triple A and B, where $A \in \bigcup_j \beta_{a0}^j$ and $B \in \bigcup_j (\beta_{a1}^j \bigcup \beta_{b1}^j)$;

$(A^{**}7)$ The system of bijections $\{\psi_B\colon B \in \bigcup_{j \in I_{3m-2}} (\beta_{10}^j \bigcup \beta_{a0}^j \bigcup \beta_{a1}^j \bigcup \beta_{b1}^j)\}$ satisfies that if $\{x, y\} = A \bigcap B$, $x \neq y$, $A \in \bigcup_j \beta_{a0}^j$ and $B \in \bigcup_j \beta_{a1}^j$ (resp. $\bigcup_j \beta_{b1}^j$) then $\psi_A(x) = \psi_B(y)$ and $\psi_B(x) = \psi_A(y)$ (resp. $\psi_A(x) = \psi_B(x)$ and $\psi_A(y) = \psi_B(y)$)

From these conditions it can be deduced that $|\bigcup_{j \in I_{3m-2}} \beta_{10}^j| = m$ and $|\bigcup_{j \in I_{3m-2}} (\beta_{a0}^j | \bigcup \beta_{a1}^j \bigcup \beta_{b1}^j)| = 3m(m-1)$. The AD^{**}-design and its large set LAD_3 had been introduced by Lu Jiaxi [Lu 4], and he gave the following examples, which can be used to construct $LSTS$ recursively.

An $LAD_3(7)$: From $(A^{**}2)$ and $(A^{**}4)$ we see that the existence of an $LAD_3(m)$ implies the existence of an $LSTS(3m)$. Thus, we might try to construct an $LAD_3(m)$ by properly dividing the triples of each STS of an $LSTS(3m)$ to form each β_{vy} of each AD^{**}. For $m = 7$, the $LAD_3(7)$ can be obtained in such way. Let $LSTS(21) = \{(F_4^* \times Z_7, \mathcal{A}_j)\colon j \in Z_7\} \bigcup \{(F_4^* \times Z_7, \mathcal{B}_{jk})\colon j \in Z_7 \backslash \{0\}, k \in F_2\}$, which is given by Theorem 2.3.1. Each $\mathcal{A}_j (j \in Z_7)$ can be divided into the following parts (where $i \in I_3, s \in \{1, 2, 4\}$):

$\beta_{10}^j : \{(1, j), (a, j), (b, j)\}$;

$\beta_{11}^j : \{(a^i, j), (a^i, j+s), (a^i, j-s)\}$; $\{(a^i, j), (a^{i+1}, j+s), (a^{i+1}, j+5s)\}$, $\{(a^i, j), (a^{i+2}, j+5s), (a^{i+2}, j+4s)\}$, $\{(a^i, j+4s), (a^{i+1}, j+4s), (a^{i+2}, j+s)\}$, $\{(a^i, j+3s), (a^{i+1}, j+3s), (a^{i+2}, j+2s)\}$, $\{(a^i, j+s), (a^{i+1}, j+$

$3s), (a^{i+2}, j+6s)\}$, $\{(a^i, j+s), (a^{i+1}, j+6s), (a^{i+2}, j+3s)\}$;

$\beta^j_{a0} : \{(a^i, 1+j), (a^i, 2+j), (a^i, 4+j)\}$;

$\beta^j_{a1} : \{(a^i, 6+j), (a^i, 5+j), (a^i, 3+j)\}$;

$\beta^j_{b1} = \phi$.

Each $\mathcal{B}_{j0}(j \in Z_7 \backslash \{0\})$ can be divided as follows (where $s \in \{1,2,4\}, t \in Z_7$):

$\beta^{j+6}_{11} : \{(1, 4j+t), (a, s+t), (a, 5s+t)\}$ $\{(a,t), (b, 6j+s+t), (b, 6j+5s+t)\}$, $\{(b, 6j+t), (1, 4j+s+t), (1, 4j+5s+t)\}$;

$\beta^{j+6}_{a0} : \{(1, 4j+t), (a,t), (b, 6j+t)\}$;

$\beta^{j+6}_{10} = \beta^{j+6}_{a1} = \beta^{j+6}_{b1} = \phi$.

Each $\mathcal{B}_{j1}(j \in Z_7 \backslash \{0\})$ can be divided as follows (where $s \in \{1,2,4\}, t \in Z_7$):

$\beta^{j+12}_{11} : \{(1, 2j+t), (b, 6j+3s+t), (b, 6j+s+t)\}$ $\{(a,t), (1, 2j+3s+t), (1, 2j+s+t)\}$, $\{(b, 6j+t), (a, 3s+t), (a, s+t)\}$;

$\beta^{j+12}_{b1} : \{(1, 2j+t), (a,t), (b, 6j+t)\}$;

$\beta^{j+12}_{10} = \beta^{j+12}_{a0} = \beta^{j+12}_{a1} = \phi$.

For each triple $B = \{x_0, x_1, x_2\}$, in the same order as listed above, take $\psi_B(x_i) = a_i, i \in I_3$, then it is easy to verify that conditions $(A^{**}1)-(A^{**}7)$ are satisfied, thereby it is an $LAD_3(7)$.

An $LAD_3(11)$:

Step 1. Construct a 3-(16,4,1) design $(F_4 \times F_4, \mathcal{B})$, where

$$\mathcal{B} = \{\{(u_0,u'_0),(u_1,u'_1),(u_2,u'_2),(u_3,u'_3)\}: \sum_{i=0}^{3} u_i = \sum_{i=0}^{3} u'_i = 0 \text{ and } (u_i,u'_i) \neq (u_j,u'_j)\ \forall i \neq j\}.$$

Set up a system of bijections $\psi_B : B \to F_4$ for $B \in \mathcal{B}$, such that $\psi_B(0,0) = 0$ if $(0,0) \in B \in \mathcal{B}$. The 3-design includes a $RB[4,1;16] = (F_4 \times F_4, \mathcal{B}')$ as a subsystem, and $\mathcal{B}'$ consists of five parallel classes η_x $(x \in F_4 \bigcup \{\infty\})$, where

$\eta_\infty = \{\{(u',0),(u',1),(u',a),(u',b)\}: u' \in F_4\}$ and

$\eta_u = \{\{(0,u'),(1,u+u'),(a,bu+u'),(b,au+u')\}: u' \in F_4\}$, $u \in F_4$.

Step 2. Construct 15 $B[4,1;16] = (F_4 \times F_4, \eta_{xv})$, $x \in F_4 \bigcup \{\infty\}$, $v \in F_4^*$.

Each η_{xv} consists of three pairwise disjoint parts: η_x, η_{xv0} and η_{xv1}, where η_x is from Step 1 and $\{\eta_{xvy}: x \in F_4 \bigcup \infty, v \in F_4\} = \mathcal{B} \backslash \mathcal{B}'$ for $y \in F_2$.

The concrete construction is as follows (where $v_1, v_2 \in F_4^*$ and u, u' run over F_4, v' runs over F_4^*):

$\eta_{\infty v0}: \{(0,u),(v',u),(av',vv'+u),(bv',vv'+u)\}$ and
$\{(0,u),(1,v+u),(a,av+u),(b,bv+u)\}$;

$\eta_{\infty v1} = \eta_{0v0} = \phi$;

$\eta_{0v1}: \{(u',u),(u',vv'+u),(v'+u',avv'+u),(v'+u',bvv'+u)\}$ and
$\{(0,u),(1,v+u),(a,av+u),(b,bv+u)\}$;

$\eta_{v_1v_20}: \{(u',u),(u',v_1v_2^2+u),(v_2+u',av_1v_2^2+u),(v_2+u',bv_1v_2^2+u)\}$ and $\{(u',u),(u',av_1v_2^2+u),(av_2+u',u),(av_2+u',av_1v_2^2+u)\}$;

$\eta_{v_1v_21}: \{(u',u),(u',bv_1v_2^2+u),(bv_2+u',u),(bv_2+u',bv_1v_2^2+u)\}$ and
$\{(0,u),(v_2,u),(av_2,v_1v_2^2+u),(bv_2,v_1v_2^2+u)\}$.

Step 3. Construct an $LSTS(9) = \{(\{\infty\}\bigcup(F_4\times F_2),\ C_x) :\ x \in \{\infty\}\bigcup(F_4^*\times F_2)\}$. The concrete constrution is as follows (where $(v,y) \in F_4^*\times F_2$ and u, v' and y' run over F_4, F_4^* and F_2, respectively):

C_∞: $\{\infty,(u,0),(u,1)\}$, $\{(1,y'),(a,y'),(b,y')\}$ and $\{(0,y'),(v',y'),(bv', y'+1)\}$;

C_{vy}: $\{\infty,(av,y),(bv,y+1)\}$, $\{(0,y+y'),(av(b+y'),y),(bv(a+y'),y+1)\}$, $\{\infty,(0,y+y'),(bv(b+y'),y+1)\}$, $\{(0,0),(0,1),(bv,y+1)\}$, $\{\infty,(bv,y),(v,y)\}$, $\{(0,y+y'),(bv(b+y'),y),(v(b+y'),y)\}$ and $\{(vv'v',y),(avv',y+1),(bvv',y+1)\}$.

Step 4. Construct an $STS(33) = (X,\mathcal{A})$, where $X = \{\infty\}\bigcup(F_4\times F_4\times F_2)$ and $\mathcal{A}$ consists of the following two parts:

(1) $\{\infty,(x,0),(x,1)\}$ with $x \in F_4\times F_4$;

(2) $\{(x_0,y_0),(x_1,y_1),(x_2,y_2)\}$ provided there exists $B \in \mathcal{B}'$ such that $\{x_0,x_1,x_2\}\subset B$ and $\{(\psi_B(x_0),y_0),(\psi_B(x_1),y_1),(\psi_B(x_2),y_2)\} \in C_\infty$.

Step 5. Construct thirty $STS(33) = (X,\mathcal{A}_{xvy})$, $x \in F_4\bigcup\{\infty\}, v \in F_4^*$, $y \in F_2$, where each $\mathcal{A}_{xvy}$ consists of the following six parts:

(1) $\{\infty,(x_1,y_1),(x_2,y_2)\}$ provided there exists $B \in \eta_x$ such that $\{x_1,x_2\}\subset B$ and $\{\infty,(\psi_B(x_1),y_1),(\psi_B(x_2),y_2)\} \in C_{vy}$;

(2) $\{(x_0,y_0),(x_1,y_1),(x_2,y_2)\}$ provided there exists $B \in \eta_x$ such that $\{x_0,x_1,x_2\}\subset B$ and $\{(\psi_B(x_0),y_0),(\psi_B(x_1),y_1),(\psi_B(x_2),y_2)\} \in C_{vy}$;

(3)—(6) $\{(x_0,y_0),(x_1,y_1),(x_2,y_2)\}$ provided there exist B and y' such that $\{x_0,x_1,x_2\}\subset B \in \eta_{xvy'}$ and the following additional conditions are satisfied for each part:

(3) $\psi_B(x_i) = a^i (i \in I_3), y_0 + y_1 = y$ and $y_0 + y_2 = y'$;

(4) $\psi_B(x_i) = a^i (i \in I_2), \psi_B(x_2) = 0, y_0 + y_1 = y+1$ and $y_0 + y_2 = y + y'$;

(5) $\psi_B(x_i) = a^{i+2} (i \in I_2), \psi_B(x_2) = 0, y_0 + y_1 = y' + 1$ and $y_1 + y_2 = y + y' + 1$;

(6) $\psi_B(x_i) = a^{i+1} (i \in I_2), \psi_B(x_2) = 0, y_0 + y_1 = y + y' + 1$ and $y_0 + y_2 = y'$.

Then $(X, \mathcal{A}) \bigcup \{(X, \mathcal{A}_{xvy}) \colon x \in F_4 \bigcup \{\infty\}, v \in F_4^*, y \in F_2\}$ is an $LSTS(33)$.

Step 6. Construct $LAD_3(11) = LAD_3[X] = \{AD^{**}(11)\} \bigcup \{AD^{**}_{xvy}(11) : x \in F_4 \bigcup \{\infty\}, v \in F_4^*, y \in F_2\}$, where $AD^{**}(11) = (\beta_{10}, \beta_{11}, \beta_{a0}, \beta_{a1}, \beta_{b1})$, and

$\beta_{10} : \{\infty, (0,0,0), (0,0,1)\}$;

$\{(u_0, u_0', y), (u_1, u_1', y), (u_2, u_2', y)\}$ with $\{(0,0), (u_0, u_0'), (u_1, u_1'), (u_2, u_2')\} \in \mathcal{B}'$ and $y \in F_2$.

$\beta_{11} = \mathcal{A} \backslash \beta_{10}$.

$\beta_{a0} = \beta_{a1} = \beta_{b1} = \phi$.

For each $AD^{**}_{xvy}(11) = (\beta_{10}^{xvy}, \beta_{11}^{xvy}, \beta_{a0}^{xvy}, \beta_{a1}^{xvy}, \beta_{b1}^{xvy})$, define

$\beta_{a0}^{xvy} : \{(u_0, u_0', y+y'), (u_1, u_1', y'), (u_2, u_2', y)\}$ with $B = \{(0,0), (u_0, u_0'), (u_1, u_1'), (u_2, u_2')\} \in \eta_{xvy'}$ and $\psi_B(u_i, u_i') = a^i$ $(i \in I_3)$; when $y = 0$, add the following triples—$\{\infty, (u_1, u_1', 0), (u_2, u_2', 1)\}$ provided $y' \in F_2$ and there exists $B \in \eta_x$ such that $\{(0,0), (u_1, u_1'), (u_2, u_2')\} \subset B, \psi_B(u_1, u_1') = av$ and $\psi_B(u_2, u_2') = bv$; $\{(0,0,y'), (u_1, u_1', 0), (u_2, u_2', 1)\}$ provided there exists $B \in \eta_x$ such that $\{(0,0), (u_1, u_1'), (u_2, u_2')\} \subset B$, $\psi_B(u_1, u_1') = av(b + y')$ and $\psi_B(u_2, u_2') = bv(a + y')$.

$\beta_{b1}^{xvy} : \{(u_0, u_0', 1 + y + y'), (u_1, u_1', 1 + y'), (u_2, u_2', 1 + y)\}$ with $B = \{(0,0), (u_0, u_0'), (u_1, u_1'), (u_2, u_2')\} \in \eta_{xvy'}$ and $\psi_B(u_i, u_i') = a^i$ $(i \in I_3)$; when $y = 1$, add the following triples —$\{\infty, (u_1, u_1', 1), (u_2, u_2', 0)\}$ provided there exists $B \in \eta_x$ such that $\{(0,0), (u_1, u_1'), (u_2, u_2')\} \subset B, \psi_B(u_1, u_1') = av$ and $\psi_B(u_2, u_2') = bv$; $\{(0,0,1 + y'), (u_1, u_1', 1), (u_2, u_2', 0)\}$ provided $y' \in F_2$ and there exists $B \in \eta_x$ such that $\{(0,0), (u_1, u_1'), (u_2, u_2')\} \subset B$, $\psi_B(u_1, u_1') = av(b + y')$ and $\psi_B(u_2, u_2') = bv(a + y')$.

$\beta_{11}^{xvy} = \mathcal{A}_{xvy} \backslash (\beta_{a0}^{xvy} \bigcup \beta_{b1}^{xvy})$.

$\beta_{10}^{xvy} = \beta_{a1}^{xvy} = \phi$.

Theorem 2.3.8. For odd n, if there exist both an $LAD_3(m)$ and an $LSTS(1+12n)$, then there exists an $LSTS(1+12mn)$.

Construction. Let X is a $3m$-set and

$LAD_3(m) = LAD_3[X] = \{AD_j^{**}(m)\colon j \in I_{3m-2}\}$ with the system of bijection $\{\psi_B\}$, where $AD_j^{**}(m) = (\beta_{10}^j, \beta_{11}^j, \beta_{a0}^j, \beta_{a1}^j, \beta_{b1}^j)$;

$LSTS(1+12n) = \{(\{\infty\} \bigcup (F_4^* \times F_4 \times Z_n), \mathcal{B}_{uw})\colon (u,w) \in F_4 \times Z_n\} \bigcup \{(\{\infty\} \bigcup (F_4^* \times F_4 \times Z_n), \mathcal{B}_{yuw})\colon (y,u,w) \in (F_2 \times F_4 \times Z_n) \backslash \{(0,0,0)\}\}$.

Since n is odd, thus there exists a $TD(4,n) = (F_4 \times Z_n, \mathcal{G}, \mathcal{A})$, where $\mathcal{G} = \{\{u\} \times Z_n : u \in F_4\}$. Below, the expression $B \in \overline{\mathcal{A}}$ will denote the fact that there is a block A such that $B \subset A \in \mathcal{A}$.

First, construct $4n(3m-2)$ triple systems $\mathcal{N}_{uw}^j$ on the set $S = \{\infty\} \bigcup (X \times F_4 \times Z_n)$, where $j \in I_{3m-2}$ and $(u,w) \in F_4 \times Z_n$. Each $\mathcal{N}_{uw}^j$ consists of the following seven parts:

$N_1 : \{\infty, (x_1,u_1,w_1), (x_2,u_2,w_2)\}$ provided there exist v, y, z and B such that $\{x_1, x_2\} \subset B \in \beta_{vy}^j$, $v \neq 1, \psi_B(x_2) = a\psi_B(x_1), \{(\psi_B(x_1), w_1), (\psi_B(x_2), w_2), (0, w)\} \in \overline{\mathcal{A}}$, $z \in F_2$ and $u_i = \psi_B(x_i)(z + by + a^i v + b^i) + u$ $(i = 1, 2)$.

$N_2 : \{\infty, (x_1,u_1,w_1), (x_2,u_2,w_2)\}$ provided there exists B such that $\{x_1, x_2\} \subset B \in \beta_{10}^j$, $\{\infty, (\psi_B(x_1), u_1, w_1), (\psi_B(x_2), u_2, w_2)\} \in \mathcal{B}_{uw}$.

N_3—$N_7 : \{(x_0,u_0,w_0), (x_1,u_1,w_1), (x_2,u_2,w_2)\}$, where the conditions for each part are as follows:

N_3: There is B such that $\{x_0, x_1, x_2\} \subseteq B \in \beta_{10}^j$ and $\{(\psi_B(x_0), u_0, w_0), (\psi_B(x_1), u_1, w_1), (\psi_B(x_2), u_2, w_2)\} \in \mathcal{B}_{uw}$.

N_4: $\{x_0, x_1, x_2\} \in \beta_{11}^j, u_0 + u_1 + u_2 = u$ and $w_0 + w_1 + w_2 = w$.

N_5: There exist v, w', y and B such that $\{x_0, x_1\} \subset B \in \beta_{vy}, x_0 \neq x_1 = x_2$. $\psi_B(x_1) = v\psi_B(x_0)$, $u_0 + u_1 + u_2 = u$, $w_1 \neq w_2, 2w' = w_1 + w_2$ and $\{(\psi_B(x_0), w_0), (\psi_B(x_1), w'), (0, w)\} \in \overline{\mathcal{A}}$.

N_6: There exist v, y and B such that $\{x_0, x_1\} \subset B \in \beta_{vy}, x_0 \neq x_1 = x_2$. $\psi_B(x_1) = v\psi_B(x_0)$, $w_1 = w_2$, $\{(\psi_B(x_0), w_0), (\psi_B(x_1), w_1), (0, w)\} \in \overline{\mathcal{A}}$ and one of the following conditions is satisfied:

(i) $u_0 + u_1 + u_2 + u = \psi_B(x_0)by$, $u_1 \neq u_2$ and $\{u_1, u_2\} \neq \{\psi_B(x_1)a + u, \psi_B(x_1)(b + v) + u\}$;

(ii) $u_0 = \psi_B(x_0)by + u$ and $\{u_1, u_2\} = \{\psi_B(x_1)a + u,\ \psi_B(x_1)(b +$

$v) + u\}$.

N_7: There exist v, y and z such that $B = \{x_0, x_1, x_2\} \in \beta_{vy}, v \neq 1$, $\psi_B(x_i) = a^i$, $\{(1, w_0), (a, w_1), (b, w_2), (0, w)\} \in \mathcal{A}$, $z \in F_2$ and $u_i = a^i(by + z) + u$ for $i \in I_3$.

Furthermore, construct $8n - 1$ triple systems $\mathcal{P}_{yuw}$ on the set S, where $(y, u, w) \in (F_2 \times F_4 \times Z_n) \backslash \{(0,0,0)\}$. Each $\mathcal{P}_{yuw}$ consists of the following three parts.

P_1 : $\{\infty, (x_1, u_1, w_1), (x_2, u_2, w_2)\}$ provided there is B such that $\{x_1, x_2\} \subset B \in \bigcup_j \beta_{10}^j$ and $\{\infty, (\psi_B(x_1), u_1, w_1), (\psi_B(x_2), u_2, w_2)\} \in \mathcal{B}_{yuw}$.

P_2 : $\{(x_0, u_0, w_0), (x_1, u_1, w_1), (x_2, u_2, w_2)\}$ provided there is B such that $\{x_0, x_1, x_2\} \subseteq B \in \bigcup_j \beta_{10}^j$ and $\{(\psi_B(x_0), u_0, w_0), (\psi_B(x_1), u_1, w_1), (\psi_B(x_2), u_2, w_2)\} \in \mathcal{B}_{yuw}$.

P_3: $\{(x_0, b(y+y')+z+u', w_0), (x_1, (y+y')+az+u', w_1), (x_2, a(y+y')+bz+u+u', w_2)\}$ provided $y', z \in F_2, u' \in F_4$, $\{(1, w_0), (a, w_1), (b, w+w_2)\} \in \overline{\mathcal{A}}$ and there exist j and v suth that $B = \{x_0, x_1, x_2\} \in \beta_{vy'}^j, v \neq 1$ and $\psi_B(x_i) = a^i$ $(i \in I_3)$.

Then $\{(S, \mathcal{N}_{uw}^j) : j \in I_{3m-2}, (u, w) \in F_4 \times Z_n\} \bigcup \{(S, \mathcal{P}_{yuw}) : (y, u, w) \in (F_2 \times F_4 \times Z_n) \backslash \{(0,0,0)\}\}$ will be an $LSTS(1 + 12mn)$.

Proof. It is not difficult to calculate that

$$\begin{aligned}
|\mathcal{N}_{uw}^j| &= |N_1| + |N_2 \cup N_3| + |N_4| + |N_5| + |N_6| + |N_7| \\
&= 6|\beta_{a0}^j \cup \beta_{a1}^j \cup \beta_{b1}^j| n + 2|\beta_{10}^j| n(1 + 12n) + 16|\beta_{11}^j| n^2 \\
&\quad + 24n(n-1)|\beta_{a0}^j \cup \beta_{a1}^j \cup \beta_{b1}^j| + 18n|\beta_{a0}^j \cup \beta_{a1}^j \cup \beta_{b1}^j| \\
&\quad + 2n|\beta_{a0}^j \cup \beta_{a1}^j \cup \beta_{b1}^j| \\
&= 2n(12n+1)\,(|\beta_{10}^j| + |\beta_{a0}^j \cup \beta_{a1}^j \cup \beta_{b1}^j|) + 16\tfrac{3m(m-1)}{2} n^2 \\
&= 2mn(12n+1) + 24mn^2(m-1) = \tfrac{12mn(12mn+1)}{6}.
\end{aligned}$$

$|\mathcal{P}_{yuw}| = |P_1 \cup P_2| + |P_3| = 2mn(1+12n) + 24m(m-1)n^2 = \frac{12mn(12mn+1)}{6}$.

Further work is tedious and omitted here, refer to [Lu 4] . □

Corollary 2.3.9. For odd n, if there exists an $LSTS(1 + 12n)$ then there exists an $LSTS(1 + 84n)$.

Proof. This follows from Theorem 2.3.8 and an $LAD_3(7)$. □

Corollary 2.3.10. For odd n, if there exists an $LSTS(1 + 12n)$ then there exists an $LSTS(1 + 132n)$.

Proof. This follows from Theorem 2.3.8 and an $LAD_3(11)$. □

2.4 Completion of the spectrum

Denote $D = \{v$: there exists an $LSTS(v)\}$. Let us list the following Lemmas, which have been given from Sections 2.2, 2.3 and other references.

(L1) If $v \in D(v > 2)$, then $3v \in D$. ([Teir 1])

(L2) If $v \in D(v \geq 7)$, then $2v + 1 \in D$. ([Ro 1])

(L3) If $v+2 \in D(v \neq 1, 23)$, then $3v \in D$. ([Theorem 2.2.1 and 2.3.1])

(L4) If $4v + 1 \in D$, then $12v + 1 \in D$. ([Corollary 2.3.4])

(L5) If $4v + 1 \in D$, then $24v + 1 \in D$. ([Corollary 2.3.7])

(L6) If $4v + 1 \in D$, then $60v + 1 \in D$. ([Corollary 2.3.5])

(L7) If $12v + 1 \in D$, then $84v + 1 \in D$. ([Corollary 2.3.9])

(L8) If $12v + 1 \in D$, then $132v + 1 \in D$. ([Corollary 2.3.10])

(L9) If $2v+1 \in D$ (v is odd, $v \geq 13$ and 5,7,11 $\dagger v$), then $12v+1 \in D$. ([Theorem 2.3.2])

In this section, for the sake of convenience, Theorem 2.3.1 is written in the form (L3), since $v + 2 \in D$ implies $v \equiv 1$ or 5 (mod 6) and $LD(v)$ exists except $v = 1, 23$ (see Theorem 2.2.1). And, Theorem 2.3.2 implies the Lemma (L9), because there exists a $TD(14, v)$ for odd $v \geq 13$ and none of 5,7,11 divides v by the conclusion about the existence of mutually orthogonal Latin squares (refer to [BJL]). Now, we can prove the last theorem of existence of $LSTS$.

Theorem 2.4.1. There exists an $LSTS(v)$ if and only if $v \equiv 1$ or 3 (mod 6), $v \geq 3$ and $v \neq 7$.

Proof. Firstly, $3 \in D$ is trivial and $7 \notin D$ is from [Cay]. Below, using induction on v, we shall show that for any $v > 7$ and $v \equiv 1$ or 3 (mod 6), if $v' \in D$ for each $v' < v, v \equiv 1$ or 3 (mod 6) and $v' > 7$, then $v \in D$. The proof will be given separately for the following cases which evidently exhaust all the possibilities: (1) $v \equiv 9$ (mod 18), (2) $v \equiv 7$ (mod 12), (3) $v \equiv 3$ or 15 (mod 18),(4) $v \equiv 1$ or 25 (mod 36), (5) $v \equiv 49$ (mod 72), (6) $v \equiv 13$ (mod 72) .

(1) $v = 18t + 9$. We have $6t + 3 \in D \rightarrow 3(6t + 3) \in D$ by (L1).

(2) $v = 12t + 7$. We have $6t + 3 \in D \rightarrow 2(6t + 3) + 1 \in D$ by (L2).

(3) $v = 18t+3$ or $18t+15$. We have $(6t+1)+2 \in D \rightarrow 3(6t+1) \in D$ $(t \neq 0)$ or $(6t+5)+2 \in D \rightarrow 3(6t+5) \in D$ $(t \neq 0, 3)$ by (L3). But, $3 \in D$ is trivial and $15, 69 \in D$ can be obtained from [Den 2].

(4) $v = 36t+1$ or $36t+25$. We have $4(3t)+1 \rightarrow 12(3t)+1$ and $4(3t+2)+1 \rightarrow 12(3t+2)+1$ by (L4).

(5) $v = 72t+49$. We have $4(3t+2)+1 \rightarrow 24(3t+2)+1$ by (L5).

(6) $v = 72t+13$.

(i) When $5|(6t+1)$, let $6t+1 = 30s+25$. We have $4(6s+5)+1 \rightarrow 60(6s+5)+1 = 12(30s+25)+1 = 72t+13$ by (L6);

(ii) When $7|(6t+1)$, let $6t+1 = 42s+7$. We have $12(6s+1)+1 \rightarrow 84(6s+1)+1 = 12(42s+7)+1 = 72t+13$ by (L7);

(iii) When $11|(6t+1)$, let $6t+1 = 66s+55$. We have $12(6s+5)+1 \rightarrow 132(6s+5)+1 = 12(66s+55)+1 = 72t+13$ by (L8);

(iv) When none of 5, 7, 11 divides $6t+1$, if $6t+1 \geq 13$, (i.e. $t \geq 2$) then $2(6t+1)+1 \rightarrow 12(6t+1)+1$ by (L9), or else if $t = 0$ or 1 we have $13 \in D$ (by [Den 2]) or $85 \in D$ (by subcase (ii), taking $s = 0$). □

This proof of the main theorem about the existence of $LSTS$ is shorter than Lu's proof [Lu 6] due to the using of Theorem 2.2.1 and Theorem 2.3.1. By (L1), (L2) and (L3), the great majority of orders can be solved. After using (L4), which is from [Lu 2], the remaining part (most diffcult) only is orders $v \equiv 13 \pmod{36}$. The Lemmas (L5)—(L9), which are from [Lu 4], are used to deal with these orders. It seems that the constructions in (L5)—(L9) are too complicated. A unified simple construction for $v \equiv 13 \pmod{36}$, or even better for all $v \equiv 1 \pmod{12}$, would further shorten and simplify the whole proof. Up to now, such a construction has been unable to produce, however.

Next, let us turn out attention to arbitrary λ. It is well known that the necessary conditions for the existence of a $B[3, \lambda; v]$ are $(k-1) \mid \lambda(v-1)$ and $k(k-1) \mid \lambda v(v-1)$. Usually, $B[3, \lambda; v]$ can be denoted by $S_\lambda(2, 3, v)$, which becomes $STS(v)$ when $\lambda = 1$. Let $\lambda^* = \lambda(2, 3, v) = gcd(v-2, 6)$, then the conditions above will simplify to the condition that $\lambda^*|\lambda$. A *large set* of disjoint $S_\lambda(2, 3, v)$, or $LS_\lambda(2, 3, v)$, is a partition of the set of all triples of a v-set X into simple $S_\lambda(2, 3, v)$. A $(2, 3, v)-$*decomposition* is defined to be an $LS_{\lambda^*}(2,3,v)$. Obviously, if an $S_{\lambda^*}(2,3,v)$ exists, then an $S_\lambda(2,3,v)$ exists if and only if $\lambda^*|\lambda$. Therefore, the existence of $LS_{\lambda^*}(2,3,v)$ implies the existence of $LS_\lambda(2,3,v)$ for any λ with $\lambda^*|\lambda$. The existence conclusion of $LS_{\lambda^*}(2,3,v)$ has been obtained (refer to [Teir 7]).

Theorem 2.4.2. An $(2,3,v)-$decomposition exists if and only if $v \neq 7$.

As the end of this Chapter, let us introduce a way to construct $LSTS(p^n+2)$, which is raised by the auther in [Kang 2]. Let F be a finite field containing p^n elements, where p is an odd prime number, $p \neq 3$ and n is a positive integer. Let g be a primitive element of $F^* = F\backslash\{0\}$ and take elements $\infty_1, \infty_2 \notin F$. Let $R = Z_{p^n-1}$ and $R^* = R\backslash\{0\}$. Denote $m = \frac{p^n-1}{2}$ and $g^\theta = 2(\theta \in R)$. Note that $g^m = -1$. For $\alpha, \beta \in R^*$, if $g^\alpha + g^\beta = 1$ then denote $\alpha C \beta$. When $p^n \equiv 5 \pmod 6$, all elements of $R^*\backslash\{m, \theta, -\theta\}$ can be partitioned into sextuples, each of which has the form $\{\alpha, \beta, -\alpha, \beta, m+\alpha-\beta, m+\beta-\alpha\}$, where $\alpha C \beta$. When $p^n \equiv 1 \pmod 6$, all elements of $R^*\backslash\{m, \theta, -\theta, \frac{m}{3}, -\frac{m}{3}\}$ can be partitioned into sextuples. Denote $\overline{R} = \{1, 2, ..., m-1\} \subset R^*$. For $\lambda \in R^*$ denote $\overline{\lambda} = \min\{\lambda, -\lambda\}$, which belongs to $\overline{R}$. Using the "PC—TC" method (refer to Subsection 3.2.2), we has given the following theorem.

Theorem 2.4.3. For odd prime number $p \neq 3$ and positive integer n, when $p^n \equiv 1$ (resp. -1) mod 6, if there exist two $t-$subsets $\{\beta_i\}_{i=1}^t$ and $\{\delta_i\}_{i=1}^t$ of $R^*\backslash\{m, \frac{2m}{3}, \frac{4m}{3}\}$ (resp. $R^*\backslash\{m\}$), where $t = \frac{p^n-7}{6}$ (resp. $\frac{p^n-5}{6}$), such that the following conditions are satisfied, then there exists an $LSTS(p^n+2)$.

(1) $\{\overline{\beta_i}\}_{i=1}^t$, $\{\overline{\delta_i}\}_{i=1}^t$ and $\{\overline{\beta_i - \delta_i}\}_{i=1}^t$ are mutually disjoint $t-$subsets of $\overline{R}\backslash\{\frac{2m}{3}\}$ (resp. $\overline{R}$).

(2) Let $\{\overline{\mu}\} \bigcup \{\overline{\beta_i}\}_{i=1}^t \bigcup \{\overline{\beta_i}\}_{i=1}^t \bigcup \{\overline{\beta_i - \delta_i}\}_{i=1}^t = \overline{R}\backslash\{\frac{2m}{3}\}$ (resp. $\overline{R}$), then $(p^n-1)/gcd(\overline{\mu}, p^n-1)$ is even.

(3) Let $\alpha_i C \beta_i$ and $\gamma_i C \delta_i$ $(1 \leq i \leq t)$, then all numbers in the set $\{\alpha_i - \gamma_i\}_{i=1}^t$ belong to different sextuples as mentioned above.

From these parameters given by this Theorem, we can construct an $LSTS(p^n+2) = \{(F \bigcup \{\infty_1, \infty_2\}, \mathcal{B}_x) : x \in F\}$, where $\mathcal{B}_x$ contains the following triples ($y \in F\backslash\{x\}$, $s = 1, 2$ and $1 \le i \le t$):

part 1. $\{\infty_1, \infty_2, x\}$,

part 2. $\{\infty_s, y, g^\lambda x + g^\mu y\}$,

part 3. $\{x, y, g^\theta x + g^m y\}$,

part 4. $\{y, g^{\alpha_i} x + g^{\beta_i} y, g^{\gamma_i} x + g^{\delta_i} y\}$,

part 5. $\{y, (1 - g^{\frac{2m}{3}})x + g^{\frac{2m}{3}} y, (1 - g^{\frac{4m}{3}})x + g^{\frac{4m}{3}} y\}$.

Remarks: (1) There is part 5 only if $p^n \equiv 1 \pmod 6$. Let $F\backslash\{x\} = \{y_{ij} : i \in Z_3, 1 \le j \le \frac{p^n-1}{3}\}$, where $y_{i+1,j} = (1-g^{\frac{2m}{3}})x + g^{\frac{2m}{3}} y_{ij}$. Then, the part 5 consists of the triples $\{y_{0j}, y_{1j}, y_{2j}\}$, where $1 \le j \le \frac{p^n-1}{3}$.

(2) Let $F\backslash\{x\} = \{z_{ij} : 0 \le i \le 1, 1 \le j \le \frac{p^n-1}{2}\}$, where $z_{1j} = g^\theta x + g^m z_{0j}$. Then, the part 3 consists of the triples $\{x, z_{0j}, z_{1j}\}$, where $1 \le j \le \frac{p^n-1}{2}$.

(3) Let $F\backslash\{x\} = \{w_{ij} : i \in Z_k, 1 \le j \le \frac{p^n-1}{k}\}$, where $k = (p^n - 1)/gcd(\mu, p^n - 1)$ and $w_{i+1,j} = g^\lambda x + g^\mu w_{ij}$. Then, the part 2 consists of the triples $\{\infty_1, w_{2i,j}, w_{2i+1,j}\}$ and $\{\infty_2, w_{2i+1,j}, w_{2i+2,j}\}$, where $0 \le i \le \frac{k-2}{2}, 1 \le j \le \frac{p^n-1}{2}$.

By this method, using computer, we has obtained some $LSTS(p+2)$, For example, $p = 13$, 31, 37 and 79.

Example 1. $LSTS(15) = \{(\{\infty_1, \infty_2\} \cup F_{13}, \mathcal{B}_x)\colon\ x \in GF(13)\}$, where $\mathcal{B}_x$ consists of the following triples:

(1) $\{\infty_1, \infty_2, x\}$,

(2) $\{\infty_s, y, g^3 x + g^5 y\}$,

(3) $\{x, y, gx + g^6 y\}$,

(4) $\{y, g^2 x + g^{10} y, g^6 x + gy\}$,

(5) $\{y, g^7 x + g^4 y, g^9 x + g^8 y\}$.

This construction is nonisomorphic to that one in [Den 2], since this one is not resolvable. For brevity, this construction can be written as

(2). (3, 5), (3). (1, 6), (4). (2, 10)—(6, 1), (5). (7, 4)—(9, 8).

We have another $LSTS(15)$ too, which can be written as:

(2). (6, 1), (3). (1, 6), (4). (4, 7)—(8, 9), (5). (7, 4)—(9, 8).

In what follows, we use the simple writing for other examples.

Example 2. $LSTS(33)$.

The first solution: (2). (4, 11), (3). (24, 15), (5). (28, 10)—(3, 20),

(4). (7, 21)—(14, 17), (6, 6)—(26, 22),
(2, 27)—(10, 28), (21, 7)—(5, 25).

The second solution: (2). (21, 7),
(4). (13, 12)—(20, 3), (6, 6)—(26, 22),
(4, 11)—(12, 13), (11, 4)—(25, 5).

The third solution: (2). (1, 9),
(4). (15, 24)—(22, 26), (9, 1)—(29, 23),
(5, 25)—(13, 12), (18, 16)—(2, 27).

The fourth solution: (2). (1, 9),
(4). (20, 3)—(27, 2), (6, 6)—(26, 22),
(5, 25)—(13, 12), (8, 19)—(22, 26).

Example 3. $LSTS(39)$.

The first solution: (2). (25, 17), (3). (1, 18), (5). (28, 12)—(34, 24),
(4). (8, 2)—(20, 23), (3, 14)—(10, 11), (2, 8)—(18, 1)
(4, 31)—(31, 4), (6, 30)—(23, 20).

The second solution: (2). (15, 13),
(4). (13, 15)—(25, 17), (3, 14)—(10, 11), (2, 8)—(18, 1),
(4, 31)—(31, 4), (6, 30)—(23, 20).

The third solution: (2). (15, 13),
(4). (18, 1)—(30, 6), (3, 14)—(10, 11), (9, 32)—(25, 17),
(11, 10)—(2, 8), (5, 27)—(22, 7).

Example 4. $LSTS(81)$. (2). (35, 75), (3). (4, 39),
(4). (11, 22)—(8, 40), (1, 43)—(69, 59), (2, 51)—(23, 5),
(3, 77)—(66, 41), (9, 29)—(17, 20), (15, 72)—(34, 48),
(26, 33)—(54, 31), (37, 64)—(74, 74), (13, 65)—(44, 53),
(20, 17)—(56, 28), (21, 55)—(55, 21), (40, 8)—(72, 15),
(5). (33, 26)—(46, 52).

Chapter 3

Large sets of Mendelsohn triple systems

3.1 Introduction

Let X be a set of v elements ($v \geq 3$). A *cyclic triple* from X is a collection of three ordered pairs $(x, y), (y, z), (z, x)$, where x, y, and z are distinct elements of X. We will denote the cyclic triple $\{(x, y), (y, z), (z, x)\}$ by $\langle x, y, z\rangle$, or $\langle y, z, x\rangle$, or $\langle z, x, y\rangle$. A *Mendelsohn triple system* on X is a pair $(X, \mathcal{B})$ where $\mathcal{B}$ is a collection of cyclic triples from X such that each ordered pair of distinct elements of X is covered by exactly λ cyclic triples from $\mathcal{B}$. The number $\mid X \mid= v$ is called the *order* of the Mendelsohn triple system $(X, \mathcal{B})$. For brevity, one denotes such system by $MTS(v, \lambda)$. It is easy to see that if $(X, \mathcal{B})$ is an $MTS(v, \lambda)$, then $\mid \mathcal{B} \mid= v(v-1)/3$. Thus, a necessary condition for the existence of a $MTS(v, \lambda)$ is $\lambda v(v-1) \equiv 0 \pmod 3$. When $\lambda = 1$, it becomes $v \equiv 0$ or $1 \pmod 3$. Usually, $MTS(v, 1)$ is briefly written by $MTS(v)$.

This kind of design was studied first by N. S. Mendelsohn [Men]. He called such a design a *generalized triple system* and proved that the necessary condition for the existence of an $MTS(v, 1)$ is also sufficient except if $v = 6$. J. C. Bermend [Ber1] independently studied the same design and obtained the same result for this question, i.e. the decomposition of the symmetric orinted complete graph into 3-circuits. Furthermore, it is well known that (see [ColRo]):

Theorem 3.1.1. The existence spectrum for $MTS(v, \lambda)$ is "$\lambda \equiv 0$ (mod 3)" or "$\lambda \not\equiv 0$ (mod 3), $v \not\equiv 2$ (mod 3) and $(v, \lambda) \neq (6, 1)$".

As examples, we give several $MTS(v)$ for smaller orders. From viewpoint of isomorphism, there are only one $MTS(3)$ and one $MTS(4)$:

$MTS(3) = (I_3, \mathcal{A})$, where $\mathcal{A} = \{\langle 0,1,2\rangle, \langle 2,1,0\rangle\}$;

$MTS(4) = (I_4, \mathcal{B})$, where $\mathcal{B} = \{\langle 0,1,2\rangle, \langle 0,2,3\rangle, \langle 0,3,1\rangle, \langle 1,3,2\rangle\}$.

And there exist exactly three pairwise nonisomorphic $MTS(7) = (I_7, \mathcal{C}_i)$, $i = 1, 2, 3$, which are listed as follows:

$\mathcal{C}_1$	$\mathcal{C}_2$	$\mathcal{C}_3$
$\langle 0,1,3\rangle\langle 3,1,0\rangle$	$\langle 0,3,4\rangle\langle 4,3,0\rangle$	$\langle 0,1,3\rangle\langle 2,5,4\rangle$
$\langle 1,2,4\rangle\langle 4,2,1\rangle$	$\langle 1,3,5\rangle\langle 5,3,1\rangle$	$\langle 0,2,6\rangle\langle 1,4,3\rangle$
$\langle 2,3,5\rangle\langle 5,3,2\rangle$	$\langle 2,3,6\rangle\langle 6,3,2\rangle$	$\langle 0,3,2\rangle\langle 1,5,6\rangle$
$\langle 3,4,6\rangle\langle 6,4,3\rangle$	$\langle 0,1,2\rangle\langle 4,6,5\rangle$	$\langle 0,4,5\rangle\langle 1,6,2\rangle$
$\langle 4,5,0\rangle\langle 0,5,4\rangle$	$\langle 0,2,5\rangle\langle 4,1,6\rangle$	$\langle 0,5,1\rangle\langle 3,4,6\rangle$
$\langle 5,6,1\rangle\langle 1,6,5\rangle$	$\langle 0,5,6\rangle\langle 4,2,1\rangle$	$\langle 0,6,4\rangle\langle 2,3,5\rangle$
$\langle 6,0,2\rangle\langle 2,0,6\rangle$	$\langle 0,6,1\rangle\langle 4,5,2\rangle$	$\langle 1,2,4\rangle\langle 3,6,5\rangle$

Note that the system $\mathcal{C}_1$ contains seven sub$-MTS(3)$, the system $\mathcal{C}_2$ contains three sub$-MTS(3)$, but $\mathcal{C}_3$ contains no sub$-MTS(3)$. From this view, we can assert that they are mutually nonisomorphic.

Let $C(X)$ be the set of all cyclic triples from the set X of v elements. Then $\mid C(X) \mid= v(v-1)(v-2)/3$. The following problem arises quite naturally. Given a v-set X, $\lambda v(v-1) \equiv 0, 1$ (mod 3) and $v \geq 3$, is it always possible to partition $C(X)$ into $t = \frac{v-2}{\lambda}$ subsets $\mathcal{B}_1, \mathcal{B}_2, \ldots, \mathcal{B}_t$ such that for $1 \leq i \leq t$ each $(X, \mathcal{B}_i)$ is an $MTS(v, \lambda)$? Such a collection of $MTS(v, \lambda)$'s is called a *large set of pairwise disjoint* $MTS(v, \lambda)$'s, we denote it by $LMTS(v, \lambda)$. The research work on this problem started in 1981. C. C. Lindner, Luc Teirlinck, C. J. Colbourn, M. J. Colbourn and Wu Lisheng have successively presented some articles with respect to the existence of $LMTS(v, 1)$ (for details see section 3.4). The author and his students, Chang Yanxun and Lei Jianguo, have thrown oneself into the work from 1987 to 1993 and have completely solved this problem for index $\lambda = 1$ and any λ, successively.

In this chapter a number of results of our research will be stated. The sections 3.2—3.4 deal with the case $\lambda = 1$, where the section 3.2 contains two direct constructions (Theorem 3.2.1 for all admissible odd orders and Theorem 3.2.2 for all orders $2^n + 2, n \geq 3$), the section 3.3 contains three recursive constructions (Theorems 3.3.1, 3.3.2 and 3.3.3), and in section 3.4 the expected spectrum for $LMTS(v, 1)$ will be completed, as well as some of other methods and results for $LMTS(v, 1)$ will be introduced briefly. Finally, in the section 3.5, the spectrum of $LMTS(v, \lambda)$ will be obtained.

3.2 Two directed constructions

3.2.1 The odd order case

For the odd order case, i.e. $v \equiv 1$ or 3 (mod 6), $v \geq 3$, if there is an $LSTS(v)$, a large set of disjoint Steiner triple systems of order v, then we can get $LMTS(v)$ by replacing each triple $\{x, y, z\}$ with the cyclic triples $\langle x, y, z\rangle$ and $\langle x, z, y\rangle$. It is well known that there exists an $LSTS(v)$ for any $v \equiv 1$ or 3 (mod 6), $v \neq 7$ (see Chapter 2). Thus, we only need to give a construction of $LMTS(7)$, which is easily: $\{(\{a, b\} \cup Z_5, \mathcal{B}_i) : i \in Z_5\}$ and each $\mathcal{B}_i$ consists of the following cyclic triples

$$\langle a,\ b,\ i\rangle, \langle b,\ a,\ i\rangle,$$

$$\langle a, 1+i, 3+i\rangle, \langle a, 2+i, 1+i\rangle, \langle a, 3+i, 4+i\rangle, \langle a, 4+i, 2+i\rangle,$$

$$\langle b, 3+i, 1+i\rangle, \langle b, 1+i, 2+i\rangle, \langle b, 4+i, 3+i\rangle, \langle b, 2+i, 4+i\rangle,$$

$$\langle i, 1+i, 4+i\rangle, \langle i, 2+i, 3+i\rangle, \langle i, 4+i, 1+i\rangle, \langle a, 3+i, 2+i\rangle.$$

Next, we do not use the results on $LSTS(v)$ but construct directly $LMTS(v)$ for any admissible odd order v.

Theorem 3.2.1. For $v \equiv 1$ or 3 (*mod* 6) and $v \geq 3$, there exists an $LMTS(v)$.

Construction. Let $u \equiv \pm 1 \pmod 6$ and $X = \{a, b\} \cup Z_u$, where Z_u is the residue class ring modulo u, and $|X| = v = u + 2$, $a, b \notin Z_u$. Construct u cyclic triple systems $\mathcal{B}_i$ ($i \in Z_u$) on X, where each $\mathcal{B}_i$ consists of three types of cyclic triples as follows.

(1) $\langle a, b, i\rangle, \langle b, a, i\rangle$; this gives two cyclic triples.

(2) $\langle a, x+i, -2x+i\rangle, \langle b, -2x+i, x+i\rangle$, where $x \in Z_u \backslash \{0\}$; this gives $2(u-1)$ cyclic triples.

(3) $\langle x, y, z\rangle, \langle x, z, y\rangle$, where $x, y, z \in Z_u$, $x < y < z$ and $x + y + z = 3i$; this gives $(u-1)(u-2)/3$ cyclic triples.

Then, $\{(X, \mathcal{B}_i) : i \in Z_u\}$ is an $LMTS(v)$.

Proof. First, for $u = 1$ (i.e. $v = 3$), $LMTS(3) = \{\{\langle a, b, 0\rangle, \langle b, a, 0\rangle\}\}$ is trivial. Next, let $u \geq 5$, notice $u \not\equiv 0$ modulo 2 and 3, it is easy to see that 2 and 3 are invertible in Z_u and $x + i \neq -2x + i$ ($x \in Z_u \backslash \{0\}$).

1° Each $(X, \mathcal{B}_i)$ is an $MTS(v)$:

Direct calculation shows that $\mathcal{B}_i$ contains $(u+1)(u+2)/3$ cyclic triples, just the number we expected. Therefore, we only need to show that no ordered pair of distinct elements of X is covered by more than one cyclic triple in $\mathcal{B}_i$, but that is clear immediately.

2° $\{(X, \mathcal{B}_i) : i \in Z_u\}$ is an $LMTS(v)$:

Again it suffies to verify that no cyclic triple is in more than one $\mathcal{B}_i$ (since we have the right total number of triples). This is obvious for the triples of types (1) and (3) and easy to see for those of type (2):

$$\langle a, x, y\rangle \in \mathcal{B}_{(y+2x)/3} \text{ and } \langle \mathrm{b}, \mathrm{x}, \mathrm{y}\rangle \in \mathcal{B}_{(\mathrm{x}+2\mathrm{y})/3}.$$

This completes the proof. □

In this theorem the established construction $LMTS(v)$ for all admissible odd orders v has such property: there exist two special elements a, b ($a \neq b \in X$) such that

$$\begin{aligned} \langle a, b, x\rangle \in \mathcal{B}_i &\Longleftrightarrow \langle b, a, x\rangle \in \mathcal{B}_i, \\ \langle a, x, y\rangle \in \mathcal{B}_i &\Longleftrightarrow \langle b, y, x\rangle \in \mathcal{B}_i, \end{aligned} \qquad (*)$$

where $x, y \in X \backslash \{a, b\}$. We call such $LMTS(v)$ a *symmetric large set of disjoint Mendelsohn triple systems* and denoted it by $SLMTS(v)$.

3.2.2 The orders $2^n + 2$ case

We have already discussed the designs $LMTS(v)$ for all odd orders v (of course, $v \not\equiv 5 \pmod 6$). For even order $v(v \not\equiv 2 \pmod 6)$, we can express v in the form $v = 2^n u + 2$, where n is a positive integer and $u \equiv \pm 1 \pmod 6$. If we can construct $LMTS(2^n + 2)$ for each positive integer n, then, using some recursive constructions given in the next section, the existence problem of $LMTS(v)$ for all v will be solved completely.

For $n = 1$, it is very easy to give the following $LMTS(2+2) = \{(\{a,b\} \cup I_2, \mathcal{B}_i) : i \in I_2\}$:

$$\mathcal{B}_0 = \{\langle a,b,0\rangle, \langle b,a,1\rangle, \langle a,0,1\rangle, \langle b,1,0\rangle\},$$

$$\mathcal{B}_1 = \{\langle a,b,1\rangle, \langle b,a,0\rangle, \langle a,1,0\rangle, \langle b,0,1\rangle\}.$$

But for $n = 2$ there does not exist an $LMTS(2^2+2)$, since there does not exist an $MTS(6)$. Now we will devote our attention to the construction of an $LMTS(2^n + 2)$ for every positive integer $n \geq 3$.

(I) The pair class and triple class. Let F be a finite field containing 2^n elements ($n \geq 3$). Its zero and unit elements are denoted by 0 and 1, respectively. Let g be a primitive element of $F^* = F\backslash\{0\}$, $\infty_1, \infty_2 \notin F$, and $R = Z_{2^n-1} = \{0, 1, \ldots, 2^n - 2\}$. Below, all operations are over F or R, respectively.

For $\alpha, \beta \in R^* = R\backslash\{0\}$, if $g^\alpha + g^\beta = 1$ then call α, β a *couple* and denote this by $\alpha C \beta$. We point out that

(T1) If $\alpha C \beta$ then $\alpha \neq \beta$ and $\beta C \alpha$, $(-\alpha)C(\beta - \alpha)$, $(2^t\alpha)C(2^t\beta)$ for any positive integer t.

(T2) If $\alpha C \beta$ and $\gamma C \delta$ then $(\text{ind }(\alpha + \delta, \beta + \gamma) - \beta)C(\delta - \beta)$ and $(\text{ind }(\alpha + \delta, \beta + \gamma) - \gamma)C(\alpha - \gamma)$, where $\epsilon = \text{ind }(\lambda, \mu)$ means $g^\lambda + g^\mu = g^\epsilon$.

Let $x \in F$ be given. For an ordered pair (y, z) of distinct elements in $F\backslash\{x\}$, we have $\frac{z-y}{x-y} \in F\backslash\{0,1\}$. So we can let $\frac{z-y}{x-y} = g^\alpha$, where $\alpha \in R^*$, and write $z = g^\alpha x + g^\beta y$, where $\beta \in R^*$ and $\alpha C \beta$. Then we say that the ordered pair $(y, z) = (y, g^\alpha x + g^\beta y)$ in $F\backslash\{x\}$ belongs to the pair class $\langle \alpha, \beta \rangle$ (briefly PC $\langle \alpha, \beta \rangle$ or $\langle \alpha \rangle$). It is easy to see that

Proposition 3.2.2. For a given element $x \in F$,

(1) Each ordered pair of distinct elements of $F\backslash\{x\}$ belongs to a uniquely determined PC. The total number of pair classes is $2^n - 2$. Each PC contains $2^n - 1$ distinct ordered pairs.

(2) For the cyclic triple $\langle g^\gamma x + g^\delta y, y, g^\alpha x + g^\beta y\rangle$, where $y \in F\backslash\{x\}$, α, β, $\gamma, \delta \in R^*$, $\alpha C\beta, \gamma C\delta$ and $\alpha \neq \gamma$, its three ordered pairs belong to the pair classes $\langle\alpha, \beta\rangle, \langle\text{ind}(\alpha+\delta, \beta+\gamma) - \beta, \delta - \beta\rangle$ and $\langle\gamma - \delta, -\delta\rangle$, respectively.

(3) For the cyclic triple $\langle u, y, g^\alpha x + g^\beta y\rangle$, where $u = x$ (or ∞_1, or ∞_2), $\alpha, \beta \in R^*$ and $\alpha C\beta$, only one among its three ordered pairs, namely $(y, g^\alpha x + g^\beta y)$, belongs to $PC\langle\alpha, \beta\rangle$.

For an ordered triple (u, v, w) of distinct elements in F, similar to the process above, we can also write $v = g^\alpha u + g^\beta w$, where $\alpha, \beta \in R$ and $\alpha C\beta$. We say that the ordered triple (u, v, w) belongs to $[\alpha, \beta]$. Since a cyclic triple $\langle u, v, w\rangle$ is equivalent to three ordered triples (u, v, w), (v, w, u) and (w, u, v), and $v = g^\alpha u + g^\beta w$ implies $w = g^{-\beta} v + g^{\alpha-\beta} u$ and $u = g^{\beta-\alpha} w + g^{-\alpha} v$, we shall say that the cyclic triple $\langle u, v, w\rangle$ belongs to the *triple class* $\{[\alpha, \beta], [-\beta, \alpha-\beta], [\beta-\alpha, -\alpha]\}$ (brief TC $\{\alpha, -\beta, \beta-\alpha\}$). Note that the three couples in a TC have the following property:

(T^*) The first coordinate of each couple equals to the negative of the second coordinate of another couple.

For convenience below, we call three such couples having property (T^*) a *trio*, and denote them by $\{\alpha, -\beta, \beta - \alpha\}$ (only write down their first coordinates). We point out that, if $A = \{\alpha, -\beta, \beta - \alpha\}$ is a trio, then $-A = \{-\alpha, \beta, \alpha - \beta\}, 2A = \{2\alpha, -2\beta, 2\beta - 2\alpha\}$ and $(-2)A = \{-2\alpha, 2\beta, 2\alpha - 2\beta\}$ are also. It is not difficult to verify that

Proporsition 3.2.3.

(1) The total number of trios (in R^*) is $\frac{2^n-2}{3}$ if n is odd, or $\frac{2^n-4}{3} + 2$ if n is even. When n is even, there are two so-called small trios$\{\theta, \theta, \theta\}$ and $\{2\theta, 2\theta, 2\theta\}$, where $\theta = \frac{2^n-1}{3}$. In all other cases each trio consists of three different numbers.

(2) Each cyclic triple of distinct elements in F belongs to a uniquely determined TC. Each small TC (corresponding to a small trio) con-

tains $\frac{2^n(2^n-1)}{3}$ pairwise distinct cyclic triples, and each of the other TC's contains $2^n(2^n-1)$ pairwise dstinct cyclic triples.

(3) The cyclic triple $\langle g^\gamma x+g^\delta y, y, g^\alpha x+g^\beta y\rangle$ belongs to the $TC\{[\alpha-\varepsilon,\gamma-\varepsilon],[\varepsilon-\gamma,\alpha-\gamma],[\gamma-\alpha,\varepsilon-\alpha]\}$, where $\alpha,\beta,\gamma,\delta\in R^*,\alpha C\beta,\gamma C\delta$ and ϵ=ind$(\alpha+\delta,\beta+\gamma)$. The cyclic triple $\langle x,y,g^\alpha x+g^\beta y\rangle$ belongs to the $TC\{[\alpha-\beta,-\beta],[\beta,\alpha],[-\alpha,\beta-\alpha]\}$.

Later, in (III), the 2^n disjoint Mendelsohn triple systems of the $LMTS$ (2^n+2) that we shall construct on the set $X=\{\infty_1,\infty_2\}\cup F$, are denoted by $(X,\mathcal{B}_x)$. And each $\mathcal{B}_x$ will contain $\langle\infty_1,\infty_2,x\rangle$, $\langle\infty_2,\infty_1,x\rangle$ and four types of cyclic triples listed in Table 3.1:

Table 3.1

name	form of cyclic triple	PC	TC
∞_1-triple	$\langle\infty_1,y,g^{\alpha_1}x+g^{\beta_1}y\rangle$	$\langle\alpha_1,\beta_1\rangle$	
∞_2-triple	$\langle\infty_2,y,g^{\alpha_2}x+g^{\beta_2}y\rangle$	$\langle\alpha_2,\beta_2\rangle$	
x-triple	$\langle x,y,g^\lambda x+g^\mu y\rangle$	$\langle\lambda,\mu\rangle$	$[\mu,\lambda]$ $[\lambda-\mu,-\mu]$ $[-\lambda,\mu-\lambda]$
y-triple	$\langle g^\gamma x+g^\delta y,y,g^\alpha x+g^\beta y\rangle$	$\langle\alpha,\beta\rangle$ $\langle\varepsilon-\beta,\delta-\beta\rangle$ $\langle\gamma-\delta,-\delta\rangle$	$[\alpha-\varepsilon,\gamma-\varepsilon]$ $[\varepsilon-\gamma,\alpha-\gamma]$ $[\gamma-\alpha,\varepsilon-\alpha]$

where $y\in F\backslash\{x\}$, $\alpha,\alpha_1,\alpha_2,\beta,\beta_1,\beta_2,\lambda,\mu,\gamma,\delta\in R^*$ and $\lambda C\mu$, $\alpha C\beta$, $\gamma C\delta$, $\alpha_1 C\beta_1$, $\alpha_2 C\beta_2$ and ε =ind$(\alpha+\delta,\beta+\gamma)$. Below, our main task will be to choose suitable parameters such that the PC and the TC listed in the table exactly are all possible PC and TC. Note that all y-triples will occupy $\frac{2^n-5}{3}$ TC if n is odd, or $\frac{2^n-7}{3}+2$ TC if n is even (including two small TC).

(II) The choice of parameters.

For a given TC, one can give $3(2^n-3)$ pairs (α,γ) such that the corresponding y-triples $\langle g^\gamma+g^\delta y,\ y,g^\alpha x+g^\beta y\rangle$ all belong to the given TC's, where $\alpha C\beta$ and $\gamma C\delta$. But the PC's given by these pairs usually are rather chaotic, which gave us much diffculties. However, luckily, we have found the following

Lemma 3.2.4. The three PC's of a y-triple $\langle g^\gamma x+g^\delta y,y,g^\alpha x+g^\beta y\rangle$ can form a trio (i.e. they meet the property (T^*)) if and only if one of the following holds:

(1) $\delta = \alpha$ (and $\gamma = \beta$). In this case the PC's and TC consist of the same couples: $\langle \alpha, \beta \rangle, \langle -\beta, \alpha - \beta \rangle$ and $\langle \beta - \alpha, -\alpha \rangle$.

(2) $\delta = \beta - \alpha$ (and $\gamma = -\alpha$). In this case the PC's are $\langle \alpha, \beta \rangle, \langle \beta - \alpha, -\alpha \rangle$ and $\langle -\beta, \alpha - \beta \rangle$, but its TC is $\{[-2\alpha, 2\beta - 2\alpha],\ [2\alpha - 2\beta, 2\beta],\ [2\beta, 2\alpha]\}$.

Proof. By the property (T^*) and the form of the three PC's given by proposition 3.2.2 (2), there are only the following two possibilities:

1° $-\delta = \alpha$ and $\delta - \beta = -(\gamma - \delta)$. This implies $\delta = \alpha$ and $\gamma = \beta$. Furthermore, ε=ind$(\alpha + \delta, \beta + \gamma)$=ind$(2\alpha, 2\beta) = 0$, which induces the case (1) by propostion 3.2.2(2) and 3.2.3(3).

2° $\beta = -(\gamma - \delta)$ and $\delta - \beta = -\alpha$. This implies $\delta = \beta - \alpha$ and $\gamma = -\alpha$. Furthermore, ε=ind$(\alpha + \delta, \beta + \gamma)$ =ind$(\beta, \beta - \alpha) = 2\beta - \alpha$, which induces the case (2) by the same reason. □

This Lemma is very useful for us in choosing the parameters in the Table 3.1. It gives two transformations between the trios: case (1) gives a transformation from trio $\{\alpha, -\beta, \beta - \alpha\}$ to itself; case (2) gives a transformation from trio $A = \{\alpha, -\beta, \beta - \alpha\}$ to trio $B = \{-2\alpha, 2\beta, 2(\alpha - \beta)\}$, i.e. each number in the second trio is (-2) times as large as each number in the first trio (we write $B = (-2)A$).

Now we construct a directed graph $G(R^*)$, which is named *trio incidence graph*. Its vertex set consists of all trios on R^*. There is an arc from vertex A to vertex B if and only if $B = (-2)A$ (we call B the *successor* of A). The graph has $\frac{2^n-2}{3}$ (resp. $\frac{2^n-4}{3}+2$) vertices, when n is odd (resp. even). It is easy to see that the graph is a union of disjoint directed circuits. Each number α of R^* belongs to a unique vertex denoted by $V(\alpha)$. Each vertex A of $G(R^*)$ belongs to a unique circuit denoted by $\sigma(A)$, and the length of the circuit is denoted by $\rho(A)$. We have the following

Proposition 3.2.5. Let the vertex $A \in G(R^*)$.

(1) If $(-2)^s A = A$, then $\rho(A) | s$;

(2) $\rho(A) = 1$ if and only if A is a small trio;

(3) If $-A \in \sigma(A) = \sigma$, then $\rho(A)$ is even. And for any $B \in \sigma$, $-B = (-2)^{\frac{\rho(A)}{2}} B \in \sigma$ ($\frac{\rho(A)}{2}$ is minimum). If $-A \in \tau \neq \sigma(A)$, then τ and

σ have the same length and $-B \in \tau$ for any $B \in \sigma$.

By this proposition, for a circuit $\sigma \in G(R^*)$ if there exists a vertex $A \in \sigma$ and a positive integer s such that $(-2)^s A = -A$, then call this circuit σ *selt-dual.*

Lemma 3.2.6. For any $n \geq 3$, $G(R^*)$ has a self-dual circuit $\sigma(V(\alpha))$.

(1) If there exists an odd $s > 1$ with $s|n$, then $\alpha = \frac{2^n-1}{2^s-1}$. And $\rho(V(\alpha)) = 2s$ (if $s \not\equiv 0 \bmod 3$) or 2 (if $s = 3$).

(2) If $4|n$, then $\alpha = \frac{2^n-1}{15}$ and $\rho(V(\alpha)) = 4$.

Proof. (1) First, we have $(2^s - 1)\alpha = 0 \pmod{2^n - 1}$, thus $(-2)^s\alpha = -2^s\alpha = -\alpha$. Thereby $(-2)^s A = -A (A = V(\alpha))$, i.e. $\sigma(A)$ is self-dual.

Since $[(-2)^{2s} - 1]\alpha = (2^{2s} - 1)\alpha = (2^s + 1)(2^n - 1) \equiv 0 \pmod{2^n - 1}$, thus $(-2)^{2s}A = A$. Thereby, $\rho(A)|2s$ and $\rho(A)$ is even by proposition 3.2.5.

When $s \not\equiv 0 \pmod 3$, let $\rho(A) = 2t$, then $t|s$. By $A = (-2)^{2t}A = 2^{2t}A$ and $\alpha \in A$, we have the following three cases (where $\alpha C \beta$):

i) $2^{2t}\alpha = \alpha$. Then $\frac{2^{2t}-1}{2^s-1}(2^n - 1) = (2^{2t} - 1)\alpha \equiv 0 \pmod{2^n - 1}$ implies $(2^s - 1)|(2^{2t} - 1)$ and $s|2t$. As s is odd, $s|t$, so $t = s$.

ii) $2^{2t}\alpha = -\beta$. Then $2^{2t}(-\beta) = \beta - \alpha$, $2^{2t}(\beta - \alpha) = \alpha$ by (T1). Thus $(2^{6t} - 1)\alpha = 0$ and $s|6t$. But 2 and 3 not divide s, hence $s|t$, which implies $t = s$.

iii) $2^{2t}\alpha = \beta - \alpha$. Then $2^{2t}(-\beta) = \alpha$, $2^{2t}(\beta - \alpha) = -\beta$ by (T1). The argument is the same as in ii).

Thereby, when $s \not\equiv 0 \pmod 3$, we always have $\rho(A) = 2s$.

When $s = 3$ we have $g^{8\alpha} = g^\alpha$. Let $x = g^\alpha$, then

$$x^8 - x = x(x-1)(x^3 + x + 1)(x^3 + x^2 + 1) = 0.$$

But $x = g^\alpha \neq 0, 1$, so $g^{3\alpha} + g^\alpha = 1$ (thus $\beta = 3\alpha$, i.e. $(-2)\alpha = -(\beta - \alpha)$) or $g^{3\alpha} + g^{2\alpha} = 1$ (thus $2\beta = 3\alpha$, i.e. $(-2)(\beta - \alpha) = -\alpha$). Both cases give $(-2)A = -A$, thus $(-2)^2 A = A$, $\rho(A) = 2$.

(2) Since $(2^4 - 1)\alpha \equiv 0 \pmod{2^n - 1}$, we have $g^{16\alpha} = g^\alpha$. Let $x = g^\alpha$, then

$$x^{16} - x = x(x^5 - 1)(x^2 + x + 1)(x^4 + x + 1)(x^4 + x^3 + 1) = 0.$$

But $g^\alpha \neq 0$, $g^{5\alpha} \neq 1$ and $g^{2\alpha}+g^\alpha+1 \neq 0$ (else $g^{3\alpha}=1$, but $3\alpha = \frac{2^n-1}{5} \equiv 0 \pmod{2^n-1}$), so $g^{4\alpha}+g^\alpha+1=0$ (thus $\beta = 4\alpha$, i.e. $(-2)^2\alpha = \beta$) or $g^{4\alpha}+g^{3\alpha}+1=0$ (thus $4\beta = 3\alpha$, i.e. $(-2)^2(\beta-\alpha) = -\alpha$). Both give $(-2)^2A = -A$. Obviously, $(-2)A \neq -A$, so $\rho(A) = 4$. □

By this lemma we can get a self-dual circuit in $G(R^*)$ for any $n \geq 3$ (if there exists an odd $s > 1$, $s|n$, then by (1); else $n = 2^k \geq 3$ and we must have $4|n$, by (2)). Here, we give a table for $3 \leq n \leq 13$.

Table 3.2

n	3	4	5	6	7	8	9	9	10	11	12	12	13
s	3	4	5	3	7	4	3	9	5	11	3	4	13
α	1	1	1	9	1	17	7371	33	1	585	273	1	
$\rho(V(\alpha))$	2	4	10	2	14	4	2	18	10	22	2	4	26

(III) The construction and examples.

Let α be given by Lemma 3.2.6, $\rho(V(\alpha)) = 2s$. Define $\alpha_i = (-2)^i\alpha$, $0 \leq i \leq s-1$, and $\lambda = (-2)^2\alpha$. Take a number γ_j from each vertex $A_j \in G(R^*)\backslash S$, where

$$S = \{V(\alpha_0), V(\alpha_1), ..., V(\alpha_{s-1}), V(\lambda)\} \bigcup S_0, \; 1 \leq j \leq t,$$

and $S_0 = \phi$ (if n odd) or $\{\{\theta,\theta,\theta\},\{2\theta,2\theta,2\theta\}\}$ (if n even, $\theta = \frac{1}{3}(2^n-1)$); $t = \frac{1}{3}(2^n-2)-s-1$ (if n odd) or $\frac{1}{3}(2^n-4)-s-1$ (if n even). Let $\alpha_i C \beta_i$, $\lambda C \mu$, $\gamma_j C \delta_j$, $0 \leq i \leq s-1$, $1 \leq j \leq t$. Then, as announced in Table 3.1, $\mathcal{B}_x$ will contain the following triples:

Part 1. $\langle \infty_1, \infty_2, x\rangle, \langle \infty_2, \infty_1, x\rangle$, this gives two blocks;

Part 2. $\langle \infty_1, y, g^{-\mu}x+g^{\lambda-\mu}y\rangle, \langle \infty_2, y, g^{\mu-\lambda}x+g^{-\lambda}y\rangle$, this gives $2(2^n-1)$ blocks;

Part 3. $\langle x, y, g^\lambda x + g^\mu y\rangle$, this gives 2^n-1 blocks;

Part 4. $\langle g^{-\alpha_i}x + g^{\beta_i-\alpha_i}y, y, g^{\alpha_i}x+g^{\beta_i}y\rangle, 0 \leq i \leq s-1$, this gives $2(2^n-1)$ blocks;

Part 5. $\langle g^{\delta_j}x + g^{\gamma_j}y, y, g^{\gamma_j}x + g^{\delta_j}y\rangle, 1 \leq j \leq t$, this gives $t(2^n-1)$ blocks;

Part 6. $\langle g^{2\theta}x+g^\theta y, y, g^\theta x+g^{2\theta}y\rangle, \langle g^\theta x+g^{2\theta}y, y, g^{2\theta}x+g^\theta y\rangle$, this gives $\frac{2}{3}(2^n-1)$ blocks,

where y runs over $F\backslash\{x\}$. Part (6) occurs only if n is even, and when y

runs over $F\backslash\{x\}$ each element produces three times but we only retain one copy of each triple.

Theorem 3.2.7. For $n \geq 3$, the above construction $\{(\{\infty_1, \infty_2\} \bigcup F, \mathcal{B}_x);\ x \in F\}$ gives an $LMTS(2^n+2)$, where $F = GF(2^n)$.

Proof. We only need to show the following three points.

1° The total number of cyclic triples in each $\mathcal{B}_x$ equals

$$2+3(2^n-1)+s(2^n-1)+(\tfrac{2^n-2}{3}-s-1)(2^n-1) \qquad (n \text{ odd})$$
$$2+3(2^n-1)+s(2^n-1)+(\tfrac{2^n-4}{3}-s-1)(2^n-1)+\tfrac{2(2^n-1)}{3} \qquad (n \text{ even}).$$

Both totals are $\frac{1}{3}(2^n+2)(2^n+1)$, exactly as expected.

2° The parts 2—6 contain all PC's.

Suppose the vertex $V(\alpha)$ in the circuit σ. By Lemma 3.2.6, σ is self-dual and $\{\alpha, -\beta, \beta-\alpha\} = -\{\lambda, -\mu, \mu-\lambda\}$. Obviously, the part 2 and part 3 contain the three PC's corresponding to the trio $V(\lambda)$. And, by Lemma 3.2.4(2), each cyclic triple in part 4 contains the three PC's corresponding to the trio $V(\alpha_i), 0 \leq i \leq s-1$. All other PC's consisting of all trios A_j and small trios, are covered by part 5 and part 6, by Lemma 3.2.4(1).

3° The parts 3—6 cover all TC's.

By Lemma 3.2.4(2), the cyclic triple in part 4 corresponding to the trio $V(\alpha_i)$ occupies the TC corresponding to the trio $V(\alpha_{i+1})$, where $0 \leq i \leq s-1$ and $V(\alpha_s) = V(\lambda)$. And the TC corresponding to $V(\alpha) = V(\alpha_0)$ is occupied by $x-$triple (in part 3). All other TC's (corresponding to trios A_j and to small trios) are occupied by the $y-$triples in part 5 and part 6 by Lemma 3.2.4(1). □

Exmple 1. $n = 3, g^3+g+1 = 0$,

$1C3, 2C6, 4C5$.

$G(Z_7^*)$: $\{1,4,2\} \rightleftharpoons \{3,6,5\}$.

Take $\alpha = \alpha_0 = 1$ (then $\beta_0 = 3, s = 1$), $\lambda = 5$ (then $\mu = 4$).

$LMTS(10) = \{(\{\infty_1, \infty_2\} \bigcup F_8, \mathcal{B}_x) :\ x \in F_8\}$, where $\mathcal{B}_x$ contains:

(1)	$\langle\infty_1,\infty_2,x\rangle, \langle\infty_2,\infty_1,x\rangle$		
(2)	$\langle\infty_1, y, g^3x+gy\rangle$	$\langle 3,1\rangle$	
	$\langle\infty_2, y, g^6x+g^2y\rangle$	$\langle 6,2\rangle$	
(3)	$\langle x, y, g^5x+g^4y\rangle$	$\langle 5,4\rangle$	$\{1,4,2\}$
(4)	$\langle g^6x+g^2y, y, gx+g^3y\rangle$	$\langle 1,3\rangle\langle 2,6\rangle\langle 4,5\rangle$	$\{3,6,5\}$

Example 2. $n = 4, g^4 + g + 1 = 0$,
$1C4, 2C8, 3C14, 5C10, 6C13, 7C9, 11C12$.

$$\begin{aligned} G(Z_{15}^*):\ & \{5,5,5\} \to \{5,5,5\}, \\ & \{10,10,10\} \to \{10,10,10\}, \\ & \{1,11,3\} \to \{8,13,9\} \to \{4,14,12\} \to \{2,7,6\} \to \{1,11,3\}. \end{aligned}$$

Take $\alpha = \alpha_0 = 1$ (then $\beta_0 = 4, s = 2$), $\alpha_1 = 13$ (then $\beta_1 = 6$), $\lambda = 4$ (then $\mu = 1$), $\gamma_1 = 2$ (then $\delta_1 = 8$).

$LMTS(18) = \{(\{\infty_1, \infty_2\} \bigcup F_{16}, \mathcal{B}_x) : x \in F_{16}\}$, where $\mathcal{B}_x$ contains:

(1)	$\langle\infty_1, \infty_2, x\rangle, \langle\infty_2, \infty_1, x\rangle$		
(2)	$\langle\infty_1, y, g^{14}x + g^3y\rangle$	$\langle 14, 3\rangle$	
	$\langle\infty_2, y, g^{12}x + g^{11}y\rangle$	$\langle 12, 11\rangle$	
(3)	$\langle x, y, g^4x + gy\rangle$	$\langle 4, 1\rangle$	$\{3, 1, 11\}$
(4)	$\langle g^{14}x + g^3y, y, gx + g^4y\rangle$	$\langle 1, 4\rangle\langle 3, 14\rangle\langle 11, 12\rangle$	$\{9, 8, 13\}$
	$\langle g^2x + g^8y, y, g^{13}x + g^6y\rangle$	$\langle 13, 6\rangle\langle 8, 2\rangle\langle 9, 7\rangle$	$\{14, 12, 4\}$
(5)	$\langle g^8x + g^2y, y, g^2x + g^8y\rangle$	$\langle 2, 8\rangle\langle 7, 9\rangle\langle 6, 13\rangle$	$\{2, 7, 6\}$
(6)	$\langle g^{10}x + g^5y, y, g^5x + g^{10}y\rangle$	$\langle 5, 10\rangle$	$\{5, 5, 5\}$
	$\langle g^5x + g^{10}y, y, g^{10}x + g^5y\rangle$	$\langle 10, 5\rangle$	$\{10, 10, 10\}$

3.3 Three recursive constructions

3.3.1 Construction $u + 2$ and $v + 2 \longrightarrow uv + 2$

Theorem 3.3.1. For $u \equiv \pm 1$ (*mod* 6) and $v \geq 3$, if there exists an $LMTS(v + 2)$ then there also exists an $LMTS(uv + 2)$.

Construction. Let $\{(\{a, b\} \cup Z_u, \mathcal{B}_i) : i \in Z_u\}$ be an $LMTS(u + 2)$, where the $\mathcal{B}_i$ is indexed such that $\langle a, b, i\rangle \in \mathcal{B}_i$. By Theorem 3.2.1, there exists such an $LMTS(u + 2)$ satisfying the so-called symmetric condition (see section 3.2.1 (*)). Let $\{(\{a, b\} \cup S_v, \mathcal{C}_j) : j \in S_v\}$ be an $LMTS(v+2)$, where $S_v = \{0, 1, \ldots, v-1\}$ is a set containing v elements, $a, b \notin Z_u \cup S_v$ and $(S_v, \circ)$ is an idempotent quasigroup of order v over S_v. Let $\alpha = (0, 1, \ldots, v-1)$ be a cyclic permutation of length v on the set S_v. Construct uv cyclic triple systems $\mathcal{T}_{ij} (i \in Z_u, j \in S_v)$ on the set $X = \{a, b\} \cup (Z_u \times S_v)$ as follows:

(1) $\langle (x,m),(y,n),(z,(m\circ n)\alpha^j)\rangle$ with $\langle x,y,z\rangle \in \mathcal{B}_i, x,y,z \in Z_u, m,n \in S_v$; this gives $v^2(u-1)(u-2)/3$ cyclic triples.

(2) $\langle (x,m),(x,n),(y,(m\circ n)\alpha^j)\rangle$ with $\langle a,x,y\rangle \in \mathcal{B}_i, x,y \in Z_u, m \neq n \in S_v$; this gives $(v^2-v)(u-1)$ cyclic triples.

(3)$\langle a,(x,m),(y,m\alpha^j)\rangle$ and $\langle b,(y,m\alpha^j),(x,m)\rangle$ with $\langle a,x,y\rangle \in \mathcal{B}_i, x,y \in Z_u, m \in S_v$; this gives $2v(u-1)$ cyclic triples.

(4)$\langle (i,m),(i,n),(i,l)\rangle$ with $\langle m,n,l\rangle \in \mathcal{C}_j$ (whenever a or b appears for m,n,l, omit the first coordinate i); this gives $(v+2)(v+1)/3$ cyclic triples.

Then, $\{(X,\mathcal{T}_{ij}); i \in Z_u, j \in S_v\}$ is an expected $LMTS(uv+2)$.

Proof. By the symmetry condition(*) of $\mathcal{B}_i$ and $\langle a,b,i\rangle \in \mathcal{B}_i$, it is easy to see that the conditions in (2) and (3) imply $x,y \in Z_u\backslash\{i\}$ and both $\langle a,x,y\rangle$ and $\langle b,y,x\rangle$ belong to $\mathcal{B}_i$.

1° Each $(\{a,b\}\bigcup(Z_n \times S_v),\mathcal{T}_{ij})$ is an $MTS(uv+2)$.

Direct calculation shows that each $\mathcal{T}_{ij}$ contains $(uv+2)(uv+1)/3$ cyclic triples, just the number we expected. Therefore we only need to show that every ordered pair P of distinct elements of the set $\{a,b\}\cup(Z_u\times S_v)$ is contained in some cyclic triple in $\mathcal{T}_{ij}$. All the possibilities are exhausted as follows:

i)$P=(a,b)\in\langle a,b,(i,m)\rangle$ in (4), where $m\in S_v$ such that $\langle a,b,m\rangle \in \mathcal{C}_j$; $P=(b,a)\in\langle b,a,(i,n)\rangle$ in (4), where $n \in S_v$ such that $\langle b,a,n\rangle \in \mathcal{C}_j$.

ii) For $x\in Z_u$ and $m \in S_v$, $P=(a,(x,m))$ belongs to the cyclic triple

$\langle b,a,(i,m)\rangle$ in (4), if $x=i$ and $\langle a,m,b\rangle \in \mathcal{C}_j$;

$\langle a,(i,m),(i,n)\rangle$ in (4), if $x=i$ and $\langle a,m,n\rangle \in \mathcal{C}_j, n\in S_v$;

$\langle a,(x,m),(y,m\alpha^j)\rangle$ in (3) if $x\neq i$ and $\langle a,x,y\rangle \in \mathcal{B}_i$.

Similarly, for $P=(b,(x,m)),((x,m),a)$ and $((x,m),b)$.

iii) For $x\in Z_u$ and $m\neq n\in S_v, P=((x,m),(x,n))$ belongs to the cyclic triple

$\langle (x,m),(x,n),(y,(m\circ n)\alpha^j)\rangle$ in (2), if $x\neq i$ and $\langle a,x,y\rangle \in \mathcal{B}_i$;

$\langle (i,m),(i,n),(i,l)\rangle$ in (4), if $x=i$ and $\langle m,n,l\rangle \in \mathcal{C}_j$.

iv) For $x\neq y\in Z_u$ and $m,n\in S_v, P=((x,m),(y,n))$ belongs to the cyclic triple

$\langle (x,m),(y,n),(z,(m\circ n)\alpha^j)\rangle$ in (1), if $\langle x,y,z\rangle \in \mathcal{B}_i$ and $z\in Z_u$;

$\langle a,(x,m),(y,m\alpha^j)\rangle$ in (3), if $\langle x,y,a\rangle \in \mathcal{B}_i$ and $m\alpha^j = n$;

$\langle (x,n'),(x,m),(y,(n'\circ m)\alpha^j)\rangle$ in (2), if $\langle x,y,a\rangle \in \mathcal{B}_i$ and $(n'\circ m)\alpha^j = n$.

$\langle b,(x,n\alpha^j),(y,n)\rangle$ in (3), if $\langle x,y,b\rangle \in \mathcal{B}_i$ and $n\alpha^j = m$;

$\langle (y,n),(y,m'),(x,(n\circ m')\alpha^j)\rangle$ in (2), if $\langle x,y,b\rangle \in \mathcal{B}_i$ and $(n\circ m')\alpha^j = m$.

2° $\{(\{a,b\}\bigcup(Z_u \times S_v),\ \mathcal{T}_{ij}) :\ i \in Z_u, j \in S_v\}$ is an $LMTS(uv+2)$.

We only need to show that every cyclic triple T from the set $\{a,b\} \cup (Z_u \times S_v)$ is contained in some $\mathcal{T}_{ij}$ mentioned above. All the possibilites are exhausted as follows:

i) $T = \langle a,b,(i,m)\rangle$, where $i \in Z_u, m \in S_v$, is contained in (4) of $\mathcal{T}_{ij}$, where $j \in S_v$ such that $\langle a,b,m\rangle \in \mathcal{C}_j$. Similarly, for $T = \langle b,a,(i,m)\rangle$.

ii) $T = \langle a,(i,m),(i,n)\rangle$, where $i \in Z_u, m \neq n \in S_v$ is contained in (4) of $\mathcal{T}_{ij}$, where $j \in S_v$, such that $\langle a,m,n\rangle \in \mathcal{C}_j$. Similarly, for $T = \langle b,(i,m),(i,n)\rangle$.

iii)$T = \langle a,(x,m),(y,n)\rangle$, where $x \neq y \in Z_u$, $m,n \in S_v$, is contained in (3) of $\mathcal{T}_{ij}$, where $i \in Z_u$, $j \in S_v$ such that $\langle a,x,y\rangle \in \mathcal{B}_i$ and $n = m\alpha^j$. Similarly, for $T = \langle b,(x,m),(y,n)\rangle$.

iv) $T = \langle (i,m),(i,n),(i,l)\rangle$, where $i \in Z_u$ and $m,n,l \in S_v$ are pairwise distinct, is contained in (4) of $\mathcal{T}_{ij}$, where $\langle m,n,l\rangle \in \mathcal{C}_j$.

v) $T = \langle (x,m),(x,n),(y,l)\rangle(x \neq y \in Z_u,\ m,n,l \in S_v, m \neq n)$ is contained in (2) of $\mathcal{T}_{ij}$, where $i \in Z_u, j \in S_v$ such that $\langle a,x,y\rangle \in \mathcal{B}_i$ and $(m\circ n)\alpha^j = l$.

vi) $T = \langle (x,m),(y,n),(z,l)\rangle(x,y,z \in Z_u, x \neq y \neq z \neq x$ and $m,n,l \in S_v)$ is contained in (1) of $\mathcal{T}_{ij}$, where $i \in Z_u, j \in S_v$ such that $\langle x,y,z\rangle \in \mathcal{B}_i$ and $(m\circ n)\alpha^j = l$.

This completes the proof. □

Theorem 3.3.2. For $u \equiv \pm 1 \pmod 6$, there exists an $LMTS(2u+2)$.

Construction. Let $\{(\{\infty_0,\infty_1\}\cup Z_u, \mathcal{C}_i) :\ i \in Z_u\}$ be an $LMTS(u+2)$, where the $\mathcal{C}_i$ is indexed such that $\langle \infty_0,\infty_1,i\rangle \in \mathcal{C}_i$. By Theorem 2.2.1 there exists such an $LMTS(u+2)$ satisfying the so-called symmetry condition (see Section 3.2.1 (*)). We can construct $2u$ cyclic triple systems $\mathcal{B}_{ij}(i \in Z_u, j \in Z_2)$ on the set $X = \{\infty_0,\infty_1\} \cup (Z_u \times Z_2)$, where the elements $(x, 0)$ and $(x, 1)$ of $Z_u \times Z_2$ are briefly denoted by x_0 and x_1.

$\mathcal{B}_{i,j}(i \in Z_u, j \in Z_2)$ consists of the following four types of triples.

(1) $\langle \infty_s, \infty_{1-s}, i_{j+s} \rangle$ and $\langle \infty_s, i_{j+s}, i_{j+s+1} \rangle$ with $s \in Z_2$; this gives four cyclic triples.

(2) $\langle \infty_0, x_t, y_{j+t} \rangle$ and $\langle \infty_1, y_{1+t}, x_{j+t} \rangle$ with $t \in Z_2$, $x, y \in Z_u$ and $\langle \infty_0, x, y \rangle \in \mathcal{C}_i$; this gives $4(u-1)$ cyclic triples.

(3) $\langle x_t, x_{1-t}, y_{t+j} \rangle$ with $t \in Z_2$, $x, y \in Z_u$ and $\langle \infty_0, x, y \rangle \in \mathcal{C}_i$; this gives $2(u-1)$ cyclic triples.

(4) $\langle x_l, y_m, z_n \rangle$ with $l, m, n \in Z_2$, $l + m + n = j$, $x, y, z \in Z_u$ and $\langle x, y, z \rangle \in \mathcal{C}_i$; this gives $4(u-1)(u-2)/3$ cyclic triples.

Then $\{(X, \mathcal{B}_{ij}) : i \in Z_u, j \in Z_2\}$ will be an expected $LMTS(2u+2)$.

Proof. First, since each $\mathcal{C}_i$ has the so-called symmetry property, both $\langle \infty_0, \infty_1, i \rangle$ and $\langle \infty_1, \infty_0, i \rangle$ belong to $\mathcal{C}_i$, and the condition in (2), (3) implies that $x, y \in Z_u \backslash \{i\}$ and both $\langle \infty_0, x, y \rangle$ and $\langle \infty_1, y, x \rangle$ belong to $\mathcal{C}_i$.

1° Each of $\mathcal{B}_{ij}$ is an $MTS(2u+2)$.

Direct calculation shows that $\mathcal{B}_{ij}$ contains $(2u+1)(2u+2)/3$ cyclic triples, just the number we expected. So, we only need to show that every ordered pair P of distinct elements of the set X is contained in some cyclic triple of $\mathcal{B}_{ij}$. All the possibilities are exhausted as follows (We only give the proof for $\mathcal{B}_{i,0}$, the proof for $\mathcal{B}_{i,1}$ is similar to it.):

i) $P = (\infty_0, \infty_1)$ and (∞_1, ∞_0) are contained in part (1).

ii) $P = (\infty_0, i_j), (\infty_1, i_j), (i_j, \infty_0)$ and (i_j, ∞_1) are contained in part (1).

iii)$P = (\infty_0, x_j), (\infty_1, x_j), (x_j, \infty_0)$ and $(x_j, \infty_1)(x \neq i)$ are contained in part (2).

iv) $P = (x_j, x_{1-j})(x \in Z_u, j \in Z_2)$ belongs to part (1), if $x = i$; or part (2), if $x \neq i$.

v) $P = (x_j, y_k)(x \neq y \in Z_u, j \in Z_2)$ belongs to

part (2)(or(3)), if $\langle x, y, \infty_0 \rangle \in \mathcal{C}_i$ and $j = k$(or $j \neq k$);

part (3)(or(2)), if $\langle x, y, \infty_1 \rangle \in \mathcal{C}_i$ and $j = k$(or $j \neq k$);

part (4), if $\langle x, y, z \rangle \in \mathcal{C}_i$ and $z \in Z_u$.

2°$\{(X, \mathcal{B}_{ij}) : i \in Z_u, j \in Z_2\}$ is an $LMTS(2u+2)$.

We only need to show that every cyclic triple T from X is contained in some $\mathcal{B}_{ij}$ mentioned above. All the possibilities are exhausted as follows:

i) $T = \langle \infty_0, \infty_1, i_j \rangle$ and $\langle \infty_0, i_j, i_{1-j} \rangle (i \in Z_u, j \in Z_2)$ are contained in (1) of $\mathcal{B}_{ij}$; $T = \langle \infty_1, \infty_0, i_j \rangle$ and $\langle \infty_1, i_j, i_{1-j} \rangle (i \in Z_u, j \in Z_2)$ are contained in (1) of $\mathcal{B}_{i,1-j}$.

ii) $T = \langle \infty_0, x_j, y_k \rangle$ (or $\langle \infty_1, x_j, y_k \rangle$) $(x \neq y \in Z_u, j, k \in Z_2)$ is contained in (2) of $\mathcal{B}_{i,s}$(or $\mathcal{B}_{i,1-s}$), where $\langle \infty_0, x, y \rangle \in \mathcal{C}_i$ and $s = j + k$.

iii) $T = \langle x_j, x_{1-j}, y_k \rangle (x \neq y \in Z_u, j, k \in Z_2)$ is contained in (3) of $\mathcal{B}_{i,0}$ or $\mathcal{B}_{i,1}$, where $\langle \infty_0, x, y \rangle \in \mathcal{C}_i$.

iv) $T = \langle x_j, y_k, z_k \rangle (x, y, z \in Z_u$ are pairwise distinct, $j, k \in Z_2)$ is contained in (4) of $\mathcal{B}_{ij}$, where $\langle x, y, z \rangle \in \mathcal{C}_i$.

This completes the proof. □

3.3.2 Construction $n + 2 \longrightarrow 4n + 2$

Let $Z_4 = \{0, 1, 2, 3\}$ and $Z_n = \{0, 1, \ldots, n-1\}$ be the residue class rings modulo 4 and n respectively, where $n > 1$ is positive integer and $n \equiv \pm 1 \pmod 6$. Take elements ∞_1 and $\infty_2 \notin Z_n$. Define

$$A_i = A_i(y, k) = c_i y + d_i k$$

and

$$Q_i = Q_i(y, k) = (i, x + A_i(y, k)),$$

where variables $x, y, k \in Z_n$, coefficients $c_i, d_i \in Z_n$, and $A_i \in Z_n, Q_i \in Z_4 \times Z_n$, $i \in Z_4$. For $t \in Z_4$, define

$$\bar{t} = \begin{cases} 0 & (t = 0,\ 2) \\ 1 & (t = 1,\ 3) \end{cases} \quad and \quad \delta = \begin{cases} 1 & (t = 0,\ 3) \\ -1 & (t = 1,\ 2) \end{cases}$$

Now, let $\{(Z_n \cup \{\infty_1, \infty_2\}, C_k) : k \in Z_n\}$ be an $LMTS(n+2)$. We can construct 4n cyclic triple systems $\mathcal{B}_k^t (k \in Z_n, t \in Z_4)$ on the set $X = (Z_4 \times Z_n) \cup \{\infty_1, \infty_2\}$ as follows:

(I) $\langle (t, u), (t, v), (t, w) \rangle$ with $\langle u, v, w \rangle \in C_k$ (whenever ∞_1 or ∞_2 appears for u,v,w, we omit the first coordinate t).

(II) $\langle Q_{t+\delta}, Q_{t+2\delta}, Q_{t+3\delta}\rangle$, where $x, y \in Z_n, y \neq k$ and $k+1$.

(III) $\langle Q_{t-\delta}, Q_{t-2\delta}, Q_{t-3\delta}\rangle$, where $x, y \in Z_n, y \neq k+1-\bar{t}$.

(IV) $\langle (j,x), (j,y), (t, \frac{x+y}{2} + A_t(k+\bar{t},k) - A_j(k+\bar{t},k))\rangle$, where $x \neq y \in Z_n$ and $j \in Z_4 \setminus \{t\}$.

(V) $\langle Q_t, Q_{t+\delta}, Q_{t+2\delta}\rangle$, $\langle Q_t, Q_{t+2\delta}, Q_{t+3\delta}\rangle$ and $\langle Q_t, Q_{t+3\delta}, Q_{t+\delta}\rangle$, where $x \in Z_n$ and $y = k + \bar{t}$.

(VI) $\langle \infty_s, Q_{t\pm1}, Q_{t\pm2}\rangle$, $\langle \infty_s, Q_{t\pm2}, Q_{t\pm3}\rangle$ and $\langle \infty_s, Q_{t\pm3}, Q_{t\pm1}\rangle$, where s=1 or 2, $x \in Z_n$ and $y = k+1-\bar{t}$. The symbol "±" in the subscript of Q takes "+" if $\bar{t}+s$ is odd or "−" if $\bar{t}+s$ is even.

Note: The appearance of denominator 2 in the part (IV) can be allowed, since $n \geq 5$ is odd, so there is an inverse of the element 2 in Z_n. Furthermore, since $n \equiv \pm1 \pmod 6$, if element x in Z_n only contains prime divisors 2 and 3, then x is invertible element in Z_n. Thus $\pm1, \pm2, \pm3$ and ±4 are especially invertible in Z_n, which will be used later.

In order to get the conditions of these $A_i(y,k)$ such that the construction mentioned above gives an $LMTS(4n+2)$, we need several lemmas below, in which the following arrays will be concerned:

$$M = [x + A_0(y,k), x + A_1(y,k), x + A_2(y,k), x + A_3(y,k)]$$

and

$$L(t,j) = [x, y, \frac{x+y}{2} + A_t(k+\bar{t},k) - A_j(k+\bar{t},k)],$$

where $t \neq j \in Z_4$ and the variables $x, y, k \in Z_n$. Each row in the arrays corresponds to the same values of x, y, and k. When x, y and k run over Z_n independently, M is an $n^3 \times 4$ array and $L(t,j)$ is an $n^3 \times 3$ array. For a fixed $k \in Z_n$, the corresponding subarrays in M and $L(t,j)$ are denoted by M_k and $L_k(t,j)$, respectively. And when $y = k$ or $k+1$, the corresponding $n^2 \times 4$ subarray is denoted by M' or M'', respectively. Similarly, the subarray with fixed k in M' or M'' is denoted by M'_k or M''_k.

Lemma 3.3.3. The $4n$ cyclic triple systems $\mathcal{B}^t_k$ given above, $k \in Z_n, t \in Z_4$, form an $LMTS(4n+2)$ if all arrays $L(t,j)$ and M satisfy the following conditions:

(S1) For fixed $k \in Z_n$, the array M_k is an OA(n, 4);

(S2) The array M is an OA(3, 4, n);

(S3) Both arrays M' and M'' are OA(n, 4). And, for fixed $k \in Z_n$, both M'_k and M''_k are disjoint $n \times 4$ subarrays of M_k;

(S4) For fixed $k \in Z_n$ and any $t \neq j \in Z_4$, the array $L_k(t,j)$ is an OA(n, 3);

(S5) For any $t \neq j \in Z_4$, the array $L(t,j)$ is an $OA(3,3,n)$.

where $OA(t,k,v)$ is an orthogonal array.

Proof. First, direct calculation shows that each $\mathcal{B}_k^t$ contains (using conditions (S1) and (S2))

$$\frac{1}{3}(n+1)(n+2)+(n^2-2n)+(n^2-n)+3(n^2-n)+3n+6n = \frac{1}{3}(4n+1)(4n+2)$$

cyclic triples, which is just the number expected. In order to prove that each $\mathcal{B}_k^t$ is an $MTS(4n+2)$, we only need to show that every ordered pair P of distinct elements of X is contained in some cyclic triple in $\mathcal{B}_k^t$. All the possibilities are exhausted as follows(where $i,j \in Z_4, u,v \in Z_n$ and s=1, 2):

(1) P=(∞_s, ∞_{3-s}) is contained in some block of part (I).

(2) P=$(\infty_s, (i,u))$ is contained in some block of part (I), when $i = t$; or else in one of part (IV). This is due to

$$\{t+1, t+2, t+3\} = \{t-1, t-2, t-3\} = Z_4 \setminus \{t\}$$

and

$$x + A_i(k+1-\bar{t}, k) = x + c_i(k+1-\bar{t}) + d_i k$$

runs over Z_n (when k,t are fixed and x runs over Z_n). This is also the case for P=$((i,u), \infty_s)$.

(3) $P = ((i,u),(i,v))$, $u \neq v$. If $i = t$ then P is contained in some block of part (I), or else in one of part (IV).

(4) $P = ((i,u),(j,v))$, $i \neq j$.

When $t \in \{i,j\}$, by (S4), P is contained in some block of part (IV) or (V). For this, we only point out that the pairs deleted from part (IV) by $x = y$ will appear in the corresponding blocks of part (V).

When $t \notin \{i,j\}$, by (S1), P is contained in some block of part (II), (III), (V) or (VI). In order to show this, we only need to note the following facts:

i) $\langle t+\delta, t+2\delta, t+3\delta\rangle$ and $\langle t-\delta, t-2\delta, t-3\delta\rangle$ are two different circuit of the three elements in $Z_4 \setminus \{t\}$.

ii) $\{k+\bar{t}, k+1-\bar{t}\} = \{k, k+1\}$.

iii) The pair deleted from part (II) and (III) by $y=k$ or $y=k+1$ will appear in some blocks in part (V) or (VI).

Finally, in order to prove that these $\mathcal{B}_k^t$ form an $LMTS(4n+2)$, it is enough to show that every cyclic triple T from the set X belongs to some $\mathcal{B}_k^t$. All the possibilities are exhausted as follows (where $i,j,t,l \in Z_4, u,v,w \in Z_n$ and $s=1,2$):

(1) $T=\langle\infty_s, \infty_{3-s}, (t,u)\rangle$ belongs to the part (I) of $\mathcal{B}_k^t$, if $\langle\infty_s, \infty_{3-s}, u\rangle \in C_k$.

(2) $T=\langle\infty_s, (i,u), (j,v)\rangle$.

If $i=j$, then T belongs to the part (I) of $\mathcal{B}_k^t$ provided $\langle\infty_s, u, v\rangle \in C_k$.

If $i \neq j$, let us list the ordered pairs formed by the first coordinates of other elements in the block contained ∞_1 in part (VI) as follows:

$$(1,2),\ (2,3) \text{ and } (3,1) \quad \text{when} \quad t=0;$$

$$(0,3),\ (3,2) \text{ and } (2,0) \quad \text{when} \quad t=1;$$

$$(3,0),\ (0,1) \text{ and } (1,3) \quad \text{when} \quad t=2;$$

$$(2,1),\ (1,0) \text{ and } (0,2) \quad \text{when} \quad t=3.$$

These pairs are just all of the pairs formed by distinct elements in Z_4. Thus, for $s=1$ and given i,j, there exists $t \in Z_4$ such that the triple class $\langle\infty_1, Q_i, Q_j\rangle$ belongs to part (VI) of $\mathcal{B}_k^t$, where the subscript k is determined by (S3), i.e. there is k satisfying $Q_i = (i, x+A_i(k+1-\bar{t}, k)) = (i,u)$ and $Q_j = (j, x+A_j(k+1-\bar{t}, k)) = (j,u)$. Similarly, for $s=2$.

(3) $T=\langle(t,u),(t,v),(t,w)\rangle, u \neq v \neq w \neq u$, belongs to part (I) of $\mathcal{B}_k^t$, if $\langle u,v,w\rangle \in C_k$.

(4) $T=\langle(j,u),(j,v),(t,w)\rangle, j\neq t$ and $u\neq v$. By (S5), for given $u\neq v$, there exists $k\in Z_n$ such that $w=\frac{u+v}{2}+A_t(k+\bar{t},k)-A_j(k+\bar{t},k)$. Therefore T belongs to the part (IV) of $\mathcal{B}_k^t$.

(5) $T=\langle(i,u),(j,v),(l,w)\rangle, i\neq j\neq l\neq i$. Table 3.3 lists the blocks in the parts (II),(III) and (V).

Table 3.3

part	$t=0$	$t=1$	$t=2$	$t=3$
(II)	$\langle Q_1,Q_2,Q_3\rangle$ $y\neq k,k+1$	$\langle Q_0,Q_3,Q_2\rangle$ $y\neq k,k+1$	$\langle Q_1,Q_0,Q_3\rangle$ $y\neq k,k+1$	$\langle Q_0,Q_1,Q_2\rangle$ $y\neq k,k+1$
(III)	$\langle Q_3,Q_2,Q_1\rangle$ $y\neq k+1$	$\langle Q_2,Q_3,Q_0\rangle$ $y\neq k$	$\langle Q_3,Q_0,Q_1\rangle$ $y\neq k+1$	$\langle Q_2,Q_1,Q_0\rangle$ $y\neq k$
(V)	$\langle Q_0,Q_1,Q_2\rangle$ $\langle Q_0,Q_2,Q_3\rangle$ $\langle Q_0,Q_3,Q_1\rangle$ $y=k$	$\langle Q_1,Q_0,Q_3\rangle$ $\langle Q_1,Q_3,Q_2\rangle$ $\langle Q_1,Q_2,Q_0\rangle$ $y=k+1$	$\langle Q_2,Q_1,Q_0\rangle$ $\langle Q_2,Q_0,Q_3\rangle$ $\langle Q_2,Q_3,Q_1\rangle$ $y=k$	$\langle Q_3,Q_0,Q_1\rangle$ $\langle Q_3,Q_1,Q_2\rangle$ $\langle Q_3,Q_2,Q_0\rangle$ $y=k+1$

It is not difficult to see that the vacant blocks in part (II) or (III) for certain t will occupy in part (V) for any other t and that any 3-circuit among Q_0,Q_1,Q_2 and Q_3 appear in part (II) or (III). By (S2), T must belong to part (II), (III) or (V) of a particular $\mathcal{B}_k^t$. □

Lemma 3.3.4. All conditions (S1)—(S5) in Lemma 3.3.3 will hold if the coefficients c_i, d_i in $A_i(y,k)=c_iy+d_ik$, $i\in Z_4$, satisfy the following conditions:

(D1) For any $i\neq j\in Z_4, c_i-c_j$ is invertible in Z_n.

(D2) For any $i\neq j\in Z_4, (c_i+d_i)-(c_j+d_j)$ is invertible in Z_n.

(D3) The value of any subdeterminant in the following 4×3 matrix R is invertible in Z_n:

$$R=\begin{pmatrix} 1 & c_0 & d_0 \\ 1 & c_1 & d_1 \\ 1 & c_2 & d_2 \\ 1 & c_3 & d_3 \end{pmatrix}.$$

Proof. Below, when (D1)—(D3) hold, the correctness of (S1)—(S5) will be verified one by one.

(S1) Given $k \in Z_n$ and two columns of the array M_k ($i \neq j \in Z_4$) for any $(u, v) \in Z_n^2$, the system of equations

$$\begin{cases} x + c_i y &= u - d_i k \\ x + c_j y &= v - d_j k \end{cases}$$

has solutions x, y in Z_n, by (D1). Namely, there are $x, y \in Z_n$ such that $Q_i = (i, u)$ and $Q_j = (j, v)$.

(S2) For any $(u, v, w) \in Z_n^3$ and any three columns i, j, l of the array M, the system of equations

$$\begin{cases} x + c_i y + d_i k = u \\ x + c_j y + d_j k = v \\ x + c_l y + d_l k = w \end{cases}$$

has solutions x, y, k in Z_n, by (D3).

(S3) For any $(u, v) \in Z_n^2$ and any two columns i, j of the array M', the system of equations

$$\begin{cases} x + (c_i + d_i)k = u \\ x + (c_j + d_j)k = v \end{cases}$$

has solutions x, k in Z_n, by (D2). This result also holds for the system of equations

$$\begin{cases} x + (c_i + d_i)k = u - c_i \\ x + (c_j + d_j)k = v - c_j \end{cases}$$

Thus, both the array M' and M'' are $OA(n, 4)$. As for the disjunction between M_k' and M_k'', we can prove a stronger conclusion, i.e. there is no common row in any two columns between M_k' and M_k''. In fact, the equation $(x + (c_i + d_i)k, x + (c_j + d_j)k) = (x' + (c_i + d_i)k + c_i, x' + (c_j + d_j)k + c_j)$ implies $c_i = c_j$, which contradicts with (S1).

(S4) Given $k \in Z_n, t \neq j \in Z_4$ and any two columns of the array $L_k(t, j)$ for any $(u, v) \in Z_n^2$, we can find $x, y \in Z_n$ such that the row corresponding to the (x, y) in the two columns is (u, v). In fact, we only need to show that the system of equations

$$\begin{cases} x & &= u \\ 2^{-1}x & +2^{-1}y &= v - ((c_t + d_t) - (c_j + d_j))k \end{cases}$$

has solutions $x, y \in Z_n$. This is obvious because the value of coefficient determinant is 2^{-1}, invertible in Z_n.

(S5) For given $t \neq j \in Z_4$ and $(u, v, w) \in Z_n^3$, there exists a row (u, v, w) in the array $L(t, j)$. In fact, the value of the coefficient determinant of the system of equations

$$\begin{cases} x & & & = u \\ & y & & = v \\ 2^{-1}x & +2^{-1}y & +((c_t + d_t) - (c_j + d_j))k & = w \end{cases}$$

is $(c_t + d_t) - (c_j + d_j)$, which is invertible in Z_n by (D2). □

Lemma 3.3.5. For $n \equiv \pm 1 \pmod 6$, $n > 1$, there exist four linear functions $A_i(y, k) = c_i y + d_i k$ ($i \in Z_4$) satisfying the conditions (D1)–(D3) in Lemma 3.3.4.

Proof. We can find eight such parameters c_i, d_i ($i \in Z_4$) such that the conditions (D1)—(D3) hold. For example, the following two groups of parameters satisfy those conditions.

	c_0	d_0	c_1	d_1	c_2	d_2	c_3	d_3
First group	0	0	−1	0	2	1	1	1
Second group	1	−1	2	−1	3	0	4	0

This choice is reasonable, since $n \geq 5$. Below, the permitted invertible elements in Z_n are $\pm 1, \pm 2, \pm 3, \pm 4$ (refer to the Note before Lemma 3.3.3).

For the first group, $c_i - c_j = \pm 1, \pm 3, \pm 1, \pm 2, \pm 2$ and ± 1; $(c_i + d_i) - (c_j + d_j) = \pm 1, \pm 4, \pm 1, \pm 3, \pm 3$ and ± 2; and the values of four subdeterminants in the matrix R are equal to $-1, -1$, 1 and 1, respectively.

For the second group, $c_i - c_j = \pm 1, \pm 1, \pm 1, \pm 2, \pm 2$ and ± 3; $(c_i + d_i) - (c_j + d_j) = \pm 1, \pm 2, \pm 1, \pm 3, \pm 3$ and ± 4; and the values of four subdeterminants in the matrix R are equal to 1, 1, -1 and -1, respectively.

□

Through Lemmas 3.3.3—3.3.5, we prove that the given construction $\{(\{\infty_1, \infty_2\} \bigcup (Z_4 \times Z_n), \mathcal{B}_k^t) : k \in Z_4, t \in Z_4\}$ is an $LMTS(4n + 2)$, indeed. Thus we have

Theorem 3.3.6. For any positive integers $n \equiv \pm 1 \pmod 6$, $n > 1$, if there exists an $LMTS(n + 2)$ then there exists an $LMTS(4n + 2)$.

Note that, in the statement of this theorem, "positive integer $n \equiv \pm 1 \pmod 6$, $n > 1$" can be restricted to mean "odd prime $n > 3$". This can be done since by using Theorem 3.3.1 and the weakened theorem, the original theorem can be achieved. For the weakened Theorem 3.3.6, in the conditions (D1)—(D3) of Lemma 3.3.5, the "invertible" can be weakened with the "nonzero" and all of the demonstrations are valid.

Now, let us choose the first group of parameters in Lemma 3.3.5 to show the concrete construction of each $\mathcal{B}_k^t$ (where x and y run over Z_n).

$\mathcal{B}_k^0$ $(k \in Z_n)$:

(I) $\langle (0,u),(0,v),(0,w)\rangle$ with $\langle u,v,w\rangle \in C_k$, whenever ∞_1 or ∞_2 appears for u, v, w, omit the first coordinate 0.

(II) $\langle (1, x-y), (2, x+2y+k), (3, x+y+k)\rangle$ with $y \neq k, k+1$.

(III) $\langle (3, x+y+k), (2, x+2y+k), (1, x-y)\rangle$ with $y \neq k+1$.

(IV) $\langle (1,x),(1,y),(0,\frac{x+y}{2}+k)\rangle$, $\langle (2,x),(2,y),(0,\frac{x+y}{2}-3k)\rangle$ and $\langle (3,x)$, $(3,y)$, $(0, \frac{x+y}{2}-2k)\rangle$ with $x \neq y$.

(V) $\langle (0,x),(1,x-k),(2,x+3k)\rangle$, $\langle (0,x),(2,x+3k),(3,x+2k)\rangle$ and $\langle (0,x),(3,x+2k),(1,x-k)\rangle$.

(VI) $\langle \infty_1,(1,x-k-1),(2,x+3k+2)\rangle$, $\langle \infty_1,(2,x+3k+2),(3,x+2k+1)\rangle$, $\langle \infty_1,(3,x+2k+1),(1,x-k-1)\rangle$, $\langle \infty_2,(3,x+2k+1),(2,x+3k+2)\rangle$, $\langle \infty_2,(2,x+3k+2),(1,x-k-1)\rangle$ and $\langle \infty_2,(1,x-k-1),(3,x+2k+1)\rangle$.

$\mathcal{B}_k^1$ $(k \in Z_n)$:

(I) $\langle (1,u),(1,v),(1,w)\rangle$ with $\langle u,v,w\rangle \in C_k$, whenever ∞_1 or ∞_2 appears for u, v, w, omit the first coordinate 1.

(II) $\langle (0,x),(3,x+y+k),(2,x+2y+k)\rangle$ with $y \neq k, k+1$.

(III) $\langle (2,x+2y+k),(3,x+y+k),(0,x)\rangle$ with $y \neq k$.

(IV) $\langle (2,x),(2,y),(1,\frac{x+y}{2}-4k-3)\rangle$, $\langle (3,x),(3,y),(1,\frac{x+y}{2}-4k-3)\rangle$ and $\langle (0,x),(0,y),(1,\frac{x+y}{2}-k-1)\rangle$ with $x \neq y$.

(V) $\langle (1,x-k-1),(0,x),(3,x+2k+1)\rangle$, $\langle (1,x-k-1),(3,x+2k+1),(2,x+3k+2)\rangle$ and $\langle (1,x-k-1),(2,x+3k+2),(0,x)\rangle$.

(VI) $\langle \infty_1,(0,x),(3,x+2k)\rangle$, $\langle \infty_1,(3,x+2k),(2,x+3k)\rangle$, $\langle \infty_1,(2,x+3k),(0,x)\rangle$, $\langle \infty_2,(2,x+3k),(3,x+2k)\rangle$, $\langle \infty_2,(3,x+2k),(0,x)\rangle$ and $\langle \infty_2,(0,x),(2,x+3k)\rangle$.

$\mathcal{B}_k^2$ $(k \in Z_n)$:

(I) $\langle(2,u),(2,v),(2,w)\rangle$ with $\langle u,v,w\rangle \in C_k$, whenever ∞_1 or ∞_2 appears for u,v,w, omit the first coordinate 2.

(II) $\langle(1,x-y),(0,x),(3,x+y+k)\rangle$ with $y \neq k, k+1$.

(III) $\langle(3,x+y+k),(0,x),(1,x-y)\rangle$ with $y \neq k+1$.

(IV) $\langle(3,x),(3,y),(2,\frac{x+y}{2}+k)\rangle$, $\langle(0,x),(0,y),(2,\frac{x+y}{2}+3k)\rangle$ and $\langle(1,x),(1,y),(2,\frac{x+y}{2}+4k)\rangle$ with $x \neq y$.

(V) $\langle(2,x+3k),(1,x-k),(0,x)\rangle$, $\langle(2,x+3k),(0,x),(3,x+2k)\rangle$ and $\langle(2,x+3k),(3,x+2k),(1,x-k)\rangle$.

(VI) $\langle\infty_1,(3,x+2k+1),(0,x)\rangle$, $\langle\infty_1,(0,x),(1,x-k-1)\rangle$, $\langle\infty_1,(1,x-k-1),(3,x+2k+1)\rangle$, $\langle\infty_2,(1,x-k-1),(0,x)\rangle$, $\langle\infty_2,(0,x),(3,x+2k+1)\rangle$ and $\langle\infty_2,(3,x+2k+1),(1,x-k-1)\rangle$.

$\mathcal{B}_k^3$ $(k \in Z_n)$:

(I) $\langle(3,u),(3,v),(3,w)\rangle$ with $\langle u,v,w\rangle \in C_k$, whenever ∞_1 or ∞_2 appears for u,v,w, omit the first coordinate 3.

(II) $\langle(0,x),(1,x-y),(2,x+2y+k)\rangle$ with $y \neq k, k+1$.

(III) $\langle(2,x+2y+k),(1,x-y),(0,x)\rangle$ with $y \neq k$.

(IV) $\langle(0,x),(0,y),(3,\frac{x+y}{2}+2k+1)\rangle$, $\langle(1,x),(1,y),(3,\frac{x+y}{2}+3k+2)\rangle$ and $\langle(2,x),(2,y),(3,\frac{x+y}{2}-k-1)\rangle$ with $x \neq y$.

(V) $\langle(3,x+2k+1),(0,x),(1,x-k-1)\rangle$, $\langle(3,x+2k+1),(1,x-k-1),(2,x+3k+2)\rangle$ and $\langle(3,x+2k+1),(2,x+3k+2),(0,x)\rangle$.

(VI) $\langle\infty_1,(2,x+3k),(1,x-k)\rangle$, $\langle\infty_1,(1,x-k),(0,x)\rangle$, $\langle\infty_1,(0,x),(2,x+3k)\rangle$, $\langle\infty_2,(0,x),(1,x-k)\rangle$, $\langle\infty_2,(1,x-k),(2,x+3k)\rangle$ and $\langle\infty_2,(2,x+3k),(0,x)\rangle$.

3.4 Spectrum of $LMTS(v,1)$

Through Sections 3.2 and 3.3, we can completely solve the existence problem of large set of pairwise disjoint Mendelsohn triple systems with $\lambda = 1$.

Theorem 3.4.1. There exists an $LMTS(v,1)$ if and only if $v \equiv 0,1 \pmod 3$, $v \geq 3$ and $v \neq 6$.

Proof. The necessity is obvious, since there exists an $MTS(v)$ only if $v \equiv 0,1 \pmod 3$, $v \geq 3$ and $v \neq 6$. Denote $v = 2^t u + 2$, where $t \geq 0$ and $u \equiv \pm 1 \pmod 6$, $u \geq 1$.

If t=0, by Theorem 3.2.1, there exists an $LMTS(v) = LMTS(u+2)$.

If t=1, by Theorem 3.3.2, there exists an $LMTS(v) = LMTS(2u+2)$.

If t=2 and $u > 1$, since there exists an $LMTS(u+2)$ by Theorem 3.2.1, there exists an $LMTS(v) = LMTS(4u+2)$ by Theorem 3.3.6. (Note that there exists no $LMTS(2^2+2)$ when $t=2$ and $u=1$.)

If $t \geq 3$, since there exists an $LMTS(u+2)$ by Theorem 3.2.1 and an $LMTS(2^t+2)$ by Theorem 3.2.7, there exists an $LMTS(v) = LMTS(2^t u+2)$ by Theorem 3.3.1. □

Below, we will introduce some of other constructing methods for $LMTS(v,1)$. Maybe, they can present nonisomorphic $LMTS(v)$ for some orders.

Firstly, in [Lin 1], C. C. Lindner gave two recursive constructions: $v \to 3v$ and $v+1 \to 3v+1$. Here we introduce his construction without proof.

Let $v \geq 3$, $v \neq 6$, and Q be an idempotent Latin square of order v on the set $Q = \{1,2,...,v\}$, which has v disjoint transversals $T_1, T_2, ..., T_v$. The Latin square $Q = (z_{ij})_{i,j\in Q}$ corresponds to a quasigroup $(Q,\circ)$, i.e. $i \circ j = z_{ij}$, where $i, j, z_{ij} \in Q$. Let $\tau = (1,2,3,...,v)$ is a cyclic permutation on Q. For each $k \in \{1,2,...,v\}$, the cell (i, x_{ik}) belongs to the transversal T_k and is occupied by y_{ik}, thus $i \circ x_{ik} = y_{ik}$. Define six permutations $\alpha_k, \beta_k, \gamma_k, \alpha_k^*, \beta_k^*$ and γ_k^* on Q as follows:

$$\begin{array}{llll}
\alpha_k: & i \mapsto x_{ik}, & \alpha_k^*: & i \mapsto y_{ik}\tau, \\
\beta_k: & x_{ik} \mapsto y_{ik}, & \beta_k^*: & y_{ik}\alpha \mapsto x_{ik}, \\
\gamma_k: & y_{ik} \mapsto i, & \gamma_k^*: & x_{ik} \mapsto i.
\end{array}$$

Denote $y_{ik}^* = y_{ik}\tau$.

Theorem 3.4.2. If there exists an $LMTS(v)$, then there exists an $LMTS(3v)$.

Construction. Define $3v-2$ cyclic triple systems on the set $X = Q \times \{1,2,3\}$ as follows:

(1) Define a collection of cyclic triples $\mathcal{A}_k (1 \leq k \leq v)$ by

(i) $\langle (i,1),(x_{ik},2),(y_{ik},3)\rangle, \langle (x_{ik},2),(i,1),(y_{ik},3)\rangle, i \in Q$; this gives $2v$ triples.

(ii) $\langle (i,1),(j,1),((i\circ j)\alpha_k,2)\rangle, \langle (i,2),(j,2),((i\circ j)\beta_k,3)\rangle, \langle (i,3),(j,3),((i\circ j)\gamma_k,1)\rangle$, $i \neq j \in Q$; this gives $3v(v-1)$ triples.

(2) Define a collection of cyclic triples $\mathcal{A}_k^*(1 \le k \le v)$ by

(i) $\langle (i,1),(x_{ik},2),(y_{ik}^*,3)\rangle, \langle (x_{ik},2),(i,1),(y_{ik}^*,3)\rangle, i \in Q$; this gives $2v$ triples.

(ii) $\langle (i,1),(j,1),((i\circ j)\alpha_k^*,3)\rangle, \langle (i,3),(j,3),((i\circ j)\beta_k^*,2)\rangle, \langle (i,2),(j,2),((i\circ j)\gamma_k^*,1)\rangle$, $i \neq j \in Q$; this gives $3v(v-1)$ triples.

(3) Let $\{(Q,\mathcal{T}_k) : 1 \le k \le v-2\}$ be an $LMTS(v)$. Define a collection $\mathcal{C}_k$ of cyclic triples by

(i) $\langle (x,i),(y,i),(z,i)\rangle$ with $\langle x,y,z\rangle \in \mathcal{T}_k, i = 1,2,3$; this gives $v(v-1)$ triples.

(ii) $\langle (i,1),(j,2),((i\circ j)\tau^{k+1},3)\rangle, \langle (j,2),(i,1),((i\circ j)\tau^{k+1},3)\rangle, i,j \in Q$; this gives $2v^2$ triples.

Then $\{(X,\mathcal{A}_k) : 1 \le k \le v\} \bigcup \{(X,\mathcal{A}_k^*) : 1 \le k \le v\} \bigcup \{(X,\mathcal{C}_k) : 1 \le k \le v-2\}$ is an $LMTS(3v)$. □

Theorem 3.4.3. If there exists an $LMTS(v+1)$ and $v \ge 3$ then there exists an $LMTS(3v+1)$.

Construction. Define $3v-1$ cyclic triple systems on the set $Y = \{\infty\} \bigcup (Q \times \{1,2,3\})$ as follows.

(1) Define a collection of cyclic triples $\mathcal{B}_k$ $(1 \le k \le v)$ by

(i) $\langle \infty,(i,1),(x_{ik},2)\rangle,, \langle \infty,(x_{ik},2),(y_{ik},3)\rangle, \langle \infty,(y_{ik},3),(i,1)\rangle, i \in Q$; this gives $3v$ triples.

(ii) $\langle (x_{ik},2),(i,1),(y_{ik},3)\rangle, i \in Q$; this gives v triples.

(iii) Same to Theorem 3.4.2 (1)(ii); this gives $3v(v-1)$ triples.

(2) Define a collection of cyclic triples $\mathcal{B}_k^*$ $(1 \le k \le v)$ by

(i) $\langle \infty,(i,1),(y_{ik}^*,3)\rangle, \langle \infty,(x_{ik},2),(i,1)\rangle, \langle \infty,(y_{ik}^*,3),(x_{ik},2)\rangle, i \in Q$; this gives $3v$ triples.

(ii) $\langle (i,1),(x_{ik},2),(y_{ik}^*,3)\rangle, i \in Q$; this gives v triples.

(iii) Same to Theorem 3.4.2 (2)(ii); this gives $3v(v-1)$ triples.

(3) Let $\{(\{\infty\} \bigcup Q, \mathcal{J}_k) : 1 \le k \le v-1\}$ be an $LMTS(v+1)$. Define a collection $\mathcal{D}_k$ of cyclic triples by

(i) $\langle (x,i),(y,i),(z,i)\rangle$ with $\langle x,y,z\rangle \in \mathcal{J}_k, i = 1,2,3$ (whenever ∞ appears for x,y,z, omit the second coordinate i); this gives $v(v+1)$ triples.

(ii) $\langle (i,1),(j,2),((i\circ j)\tau^{k+1},3)\rangle, \langle (j,2),(i,1),((i\circ j)\tau^{k},3)\rangle, i,j \in Q$; this gives $2v^2$ triples. [Note: the two exponents of τ are different.]

Then, $\{(Y,\mathcal{B}_k) : 1 \le k \le v\} \bigcup \{(Y,\mathcal{B}_k^*) : 1 \le k \le v\} \bigcup \{(y,\mathcal{D}_k) : 1 \le k \le v-1\}$ is an $LMTS(3v+1)$. □

In Subesction 3.2.1, we introduced a special kind of $LMTS$, namely the symmetric $LMTS$ $\{(X,B_i)\}_i$ defined as follows. There exist $a \ne b \in X$ such that the following conditions hold for any i and $x,y \notin \{a,b\}$:

$(S1)$ $\langle a,b,x\rangle \in B_i$ if and only if $\langle b,a,x\rangle \in B_i$;

$(S2)$ $\langle a,x,y\rangle \in B_i$ if and only if $\langle b,y,x\rangle \in B_i$.

For the convenience, call the special elements a and b the *nuclei*. In this subsection, another kind of $LMTS$ is used. This is the so-called quasi-symmetric $LMTS$, which only satisfies the condition $(S1)$ mentioned above. By known constructions we can get some examples of quasi- symmetric $LMTS$.

Lemma 3.4.4. (a) There exists a quasi-symmetric $LMTS(u+2)$ for $u \equiv 1$ and $5 \pmod 6$.

(b) There exists a quasi-symmetric $LMTS(2^n+2)$ for any positive integer $n \ge 3$.

(c) For $v \ge 3$, if there exists a symmetric $LMTS(u+2)$ and a quasi-symmetric $LMTS(v+2)$ then there exists a quasi-symmetric $LMTS(uv+2)$.

(d) For odd $v \ge 3$, if there exists a quasi-symmetric $LMTS(v)$ (resp. $LMTS(v+1)$) then there exists a quasi-symmetric $LMTS(3v)$ (resp. $LMTS(3v+1)$)

Proof. (a),(b) These are trivial by Theorem 3.2.1 and Theorem 3.2.7.

(c) In the part (4) in construction of Theorem 3.3.1, if $LMTS(v+2) = \{(\{a,b\}\bigcup S_v, C_j) : j \in S_v\}$ is quasi-symmetric, then $\langle a,b,j\rangle, \langle b,a,j\rangle \in C_j$ implies $\langle a,b,(i,j)\rangle$ and $\langle b,a,(i,j)\rangle \in \mathcal{T}_{ij}$ for any $i \in Z_u$, $j \in S_v$.

(d) First, for odd $v \ge 3$ there exists a symmetric idempotent Latin square Q of order v having v disjoint transversals. Denote the corresponding quasigroup by $(Q,\cdot)$. In the construction of Theorem 3.4.2, if the nuclei of the known quasi-symmetric $LMTS(v) = \{(Q,\mathcal{T}_k) : 1 \le k \le v-2\}$ are a and b (i.e. $\langle a,b,k\rangle, \langle b,a,k\rangle \in \mathcal{T}_k$), then we have: $\langle (a,1),(b,1),((a\cdot b)\alpha_k,2)\rangle, \langle (b,1),(a,1),((b\cdot a)\alpha_k,2)\rangle \in \mathcal{A}_k$ (part (1) (ii)); $\langle (a,1),(b,1),((a\cdot b)\alpha_k^*,3)\rangle, \langle (b,1),(a,1),((b\cdot a)\alpha_k^*,3)\rangle \in \mathcal{A}_k^*$ (part (2) (ii)); $\langle (a,1),(b,1),(k,1)\rangle, \langle (b,1),(a,1),(k,1)\rangle \in \mathcal{C}_k$ (part (3)(i)). Therefore, we can regard elements $(a,1),(b,1) \in Q\times\{1,2,3\}$ as the nuclei of the $LMTS(3v) = \{(X,\mathcal{A}_k) : 1\le k\le v\} \bigcup \{(X,\mathcal{A}_k^*) : 1\le k\le v\} \bigcup \{(X,\mathcal{C}_k) : 1 \le k \le v-2\}$. □

Remarks. (1) An example of symmetric idempotent Latin square Q of order v having v disjoint transversals is $Q = (a_{ij})_0^{v-1}$, where $a_{ij} = 2^{-1}(i+j) \pmod v$, and its orthogonal mate is $P = (b_{ij})_0^{v-1}$, where $b_{ij} = 2i+j \pmod v$. Here, v is odd and $v\ge 3$.

(2) We point out that, in construction (c), if $LMTS(v+2)$ is symmetric then $LMTS(uv+2)$ is as well. And for construction (a), the $LMTS(u+2)$ provided in Theorem 3.2.1 is symmetric.

Using quasi-symmetric $LMTS$ and $LD-$design, the following Theorem has been obtained in [KangChang 2].

Theorem 3.4.5. If an $LD(n)$ and a quasi-symmetric $LMTS(n+2)$ both exist, then an $LMTS(3n)$ exists.

Construction. Let $I_n = \{0,1,...,n-1\}$, $a,b \notin I_n$ and $Z_3 = \{0,1,2\}$. Let $LD(n) = (I_n, \mathcal{L}^1, \mathcal{L}^2, (\mathcal{L}_x)_{x\in I_n})$ with $(x,x,x,0) \in \mathcal{L}^1 \bigcap \mathcal{L}^2$ for any $x \in I_n$, and $LMTS(n+2) = \{(\{a,b\}\bigcup I_n, \mathcal{C}_x) : x\in I_n\}$ with $\langle a,b,x\rangle, \langle b,a,x\rangle \in \mathcal{C}_x$.

Define n systems $\mathcal{A}_x (x\in I_n)$ of cyclic triples on $I_n \times Z_3$ as follows, where the element $(u,k)\in I_n\times Z_3$ is denoted by u_k:

(A1) $\langle x_0,x_1,x_2\rangle, \langle x_0,x_2,x_1\rangle$;

(A2) $\langle \alpha_0,\beta_1,\gamma_2\rangle, \langle \alpha_0,\gamma_2,\beta_1\rangle$ provided $(\alpha,\beta,\gamma)\in\mathcal{L}_x$;

(A3) $\langle u_k,v_k,w_k\rangle$ provided $\langle u,v,w\rangle \in \mathcal{C}_x, u,v,w\in I_n$ and $k\in Z_3$;

(A4) $\langle x_k, y_{k+1}, z_{k+1}\rangle$ provided $\langle a,y,z\rangle \in \mathcal{C}_x, y,z\in I_n$ and $k \in Z_3$;

(A5) $\langle x_k, s_{k-1}, t_{k-1}\rangle$ provided $\langle b,s,t\rangle \in \mathcal{C}_x, s,t\in I_n$ and $k\in Z_3$.

Define $2(n-1)$ systems $\mathcal{B}_{ij} (i\in I_n\backslash\{0\}, j\in\{1,2\})$ of cyclic triples on $I_n\times Z_3$ as follows:

(B1) $\langle x_0, y_1, z_2\rangle, \langle x_0, z_2, y_1\rangle$ provided $(x,y,z,i) \in \mathcal{L}^j$;

(B2) If $(x,y,z,i) \in \mathcal{L}^j$ then the following cyclic triples belong to $\mathcal{B}_{ij}$:

when $j = 1$:
$\langle x_0, u_1, v_1\rangle$ provided $\langle a,u,v\rangle \in \mathcal{C}_y$,
$\langle y_1, u_2, v_2\rangle$ provided $\langle a,u,v\rangle \in \mathcal{C}_z$,
$\langle z_2, u_0, v_0\rangle$ provided $\langle a,u,v\rangle \in \mathcal{C}_x$;
when $j = 2$:
$\langle x_0, s_2, t_2\rangle$ provided $\langle b,s,t\rangle \in \mathcal{C}_z$,
$\langle y_1, s_0, t_0\rangle$ provided $\langle b,s,t\rangle \in \mathcal{C}_x$,
$\langle z_2, s_1, t_1\rangle$ provided $\langle b,s,t\rangle \in \mathcal{C}_y$.

Then , $\{(I_n \times Z_3, \mathcal{A}_x) : x \in I_n\} \bigcup \{(I_n \times Z_3, \mathcal{B}_{ij}) : i \in I_n \backslash \{0\}, j = 1, 2\}$ is an $LMTS(3n)$.

Proof. 1° Each $\mathcal{A}_x$ is an $MTS(3n)$.

Direct calculation shows that each $\mathcal{A}_x$ contains

$$2 + 2(n-1)^2 + 3(n-1)(n-2)/3 + 3(n-1) + 3(n-1) = 3n(3n-1)/3$$

cyclic triples, just the number we expected. So we only need to show that every ordered pair P of distinct elements of the set $I_n \times Z_3$ is contained in some cyclic triple of $\mathcal{A}_x$. All the possibilities are exhausted as follows:

(1) $p = (x_i, y_j), i \neq j \in Z_3$, is contained in (A1);

(2) $p = (x_i, u_i)$ or $(u_i, x_i), u \in I_n \backslash \{x\}, i \in Z_3$, is contained in (A3);

(3) $p = (x_i, u_j), u \in I_n \backslash \{x\}, i \neq j \in Z_3$, is contained in (A4) or (A5) if $j \equiv i+1$ or $j \equiv i-1 \pmod 3$, and similarly for $p = (u_j, x_i)$;

(4) $p = (u_i, v_j),\ u, v \in I_n \backslash \{x\}, i \neq j \in Z_3$, is contained in (A2);

(5) $p = (u_i, v_i),\ u \neq v \in I_n \backslash \{x\}, i \in Z_3$. There exists $w \in \{a,b\} \bigcup I_n$ such that $\langle u,v,w\rangle \in \mathcal{C}_x$. If $w = a$ or b then P is contained in (A4) or (A5), otherwise in (A3).

2° Each $\mathcal{B}_{ij}$ is an $MTS(3n)$.

Each $\mathcal{B}_{ij}$ contains $2n + 3n(n-1) = 3n(3n-1)/3$ cyclic triples. All ordered pairs P of the set $I_n \times Z_3$ occur as follows.

(1) The pair $p = (u_k, v_k)(u \neq v \in I_n, k \in Z_3)$ is contained in (B2). For example, consider the case $k = 0$ and $j = 1$. There exists $x, z \in I_n$, such that $\langle a,u,v\rangle \in \mathcal{C}_x$ and $(x, *, z, j) \in \mathcal{L}^1$. Then we have $\langle z_2, u_0, v_0\rangle \in \mathcal{B}_{i1}$. The remaining cases are similar.

(2) The pair $p = (u_t, u_k)(u \in I_n, t \neq k \in Z_3)$ is contained in (B2). For example, consider the case $t = 0, k = 1$ and $j = 2$ (similarly for other case of t, k and j). There exists $y \in I_n \backslash \{u\}$, such that $\langle y, u, *, i\rangle \in \mathcal{L}^2$. Furthermore, there exists $v \in I_n \backslash \{u, y\}$ such that $\langle b, v, u\rangle \in \mathcal{C}_y$. Then we have $\langle u_1, v_0, u_0\rangle \in \mathcal{C}_{i2}$.

(3) The pair $p = (u_t, v_k)(u \neq v \in I_n, t \neq k \in Z_3)$. We only consider the case $t = 2, k = 1$ and $j = 1$ (other cases are similar). There exists $z \in I_n \backslash \{v\}$ such that $\langle *, v, z, i\rangle \in \mathcal{L}^1$. If $z = u$, then P is contained in $\langle *_0, u_2, v_1\rangle$ of part (B1). If $z \neq u$, then there exists $w \in I_n \backslash \{z, u\}$ such that $\langle a, w, u\rangle \in \mathcal{C}_z$, so P is contained in $\langle v_1, w_2, u_2\rangle$ of part (B2).

3° $\{(I_n \times Z_3, \mathcal{A}_x) : x \in I_n\} \bigcup \{(I_n \times Z_3, \mathcal{B}_{ij}) : i \in I_n \backslash \{0\}, j = 1, 2\}$ is an $LMTS(3n)$.

We only need to show that every cyclic triple T from $I_n \times Z_3$ is contained in some $\mathcal{A}_x$ or $\mathcal{B}_{ij}$ mentioned above. All the possibilities are exhausted as follows:

(1) $T = \langle u_k, v_k, w_k\rangle(u \neq v \neq w \neq u \in I_n, k \in Z_3)$ is contained in (A3) of $\mathcal{A}_x$ if $\langle u, v, w\rangle \in \mathcal{C}_x$;

(2) $T = \langle u_0, v_1, w_2\rangle$ or $\langle u_0, w_2, v_1\rangle(u, v, w \in I_n)$. If $u = v = w$ then T is contained in (A1) of $\mathcal{A}_u$; in other cases, by the properties of $LD(n)$, $(u, v, w) \in \mathcal{L}_x$ or $(u, v, w, i) \in \mathcal{L}^j$, so T is contained in (A2) of $\mathcal{A}_x$ or in (B1) of $\mathcal{B}_{ij}$;

(3) $T = \langle u_t, v_k, w_k\rangle(u, v, w \in I_n, t \neq k \in Z_3, v \neq w)$ leads to two cases.

Case 1: $k \equiv t + 1 \pmod 3$, as for example $t = 0, k = 1$. There exists $x \in I_n \backslash \{v, w\}$ such that $\langle a, v, w\rangle \in \mathcal{C}_x$. If $x = u$, then T is contained in (A4) of $\mathcal{A}_x$, or else in (B2) of $\mathcal{B}_{i1}$ provided $(u, x, *, i) \in \mathcal{L}^1$. The other cases are similar.

Case 2: $k \equiv t - 1 \pmod 3$, as for example $t = 1, k = 0$. There exists $x \in I_n \backslash \{v, w\}$ such that $\langle b, v, w\rangle \in \mathcal{C}_x$. If $x = u$, then T is contained in (A5) of $\mathcal{A}_x$, or else in (B2) of $\mathcal{B}_{i2}$ provided $(x, u, *, i) \in \mathcal{L}^2$. The other cases are similar.

The proof is completed. □

For a quasi-symmetric $LMTS$, $QLMTS(v) = \{(X, \mathcal{B}_i): 1 \leq i \leq v - 2\}$ with its nuclei a and b, if there exists a subset $\mathcal{J}_i$ of $\mathcal{B}_i$, for any

$i = 1, 2, ..., v-2$, satisfying the additional conditions:

($S3$) Each $\mathcal{J}_i$ forms a partition of $X \backslash \{a, b, x_i\}$, where $\langle a, b, x_i \rangle$ and $\langle b, a, x_i \rangle \in \mathcal{B}_i$, and

($S4$) All ordered pair of cyclic triples in $\bigcup_{i=1}^{v-2} \mathcal{J}_i$ are distinct,

then the $QLMTS(v)$ is called a *compatible* $LMTS(v)$, and denoted by $CLMTS$ (v). Each $\mathcal{J}_i$ is called a ***divisible subset*** of $\mathcal{B}_i$. It is easy to know that $3|v$, and all ordered pairs of cyclic triple in $\bigcup_{i=1}^{v-2} \mathcal{J}_i$ are all pairs of distinct elements in $X \backslash \{a, b\}$. If an $CLMTS(v)$ satisfies condition ($S2$), then called it is a *compatible symmetric* $LMTS(v)$, and denoted by $CSLMTS(v)$.

Lemma 3.4.6. For prime power $q \equiv 1 \pmod 6$, there exists an $CSLMTS$ $(q+2)$.

Construction. Let $a, b \notin Z_q, X = \{a, b\} \bigcup Z_q$, construct q cyclic triple systems $\mathcal{B}_i$ on X as follows:

(1) $\langle a, b, i \rangle, \langle b, a, i \rangle$.

(2) $\langle a, x+i, -2x+i \rangle, \langle b, -2x+i, x+i \rangle$, where $x \in Z_q \backslash \{0\}$.

(3) $\langle x, y, z \rangle, \langle y, x, z \rangle$, where $\{x, y, z\}$ is unordered 3-subset of Z_q such that $x + y + z \equiv 3i \pmod q$.

Proof. In Theorem 3.2.1 we proved that $\{(X, \mathcal{B}_i)\}_{i=1}^q$ is an $SLMTS(q+2)$. Let g be a primitive element of $Z_q, m = \frac{q-1}{3}$. Define

$$\mathcal{J}_i = \{\langle g^k + i, g^{m+k} + i, g^{2m+k} + i \rangle : 0 \le k \le m-1\}.$$

Then, each $\mathcal{J}_i$ is a subset of $\mathcal{B}_i$. It is easy to check that $\mathcal{J}_i$ is a divisible subset of $\mathcal{B}_i$. So, $\{(X, \mathcal{B}_i)\}_{i=1}^q$ is an $CSLMTS(q+2)$. □

Lemma 3.4.7. If both an $CSLMTS(u+2)$ and an $CLMTS(v+2)$ exist, $v \ge 3$ is odd, then an $CLMTS(uv+2)$ exists.

Construction. Let $CSLMTS(u+2) = \{(\{a, b\} \bigcup I_u, \mathcal{B}_i) : i \in I_u\}$ with $\langle a, b, i \rangle \in \mathcal{B}_i$ and $\mathcal{J}_i$ be divisible subset of $\mathcal{B}_i$. And $CLMTS(v+2) = \{(\{a, b\} \bigcup Q, \mathcal{C}_j) : j \in Q\}$ with divisible subset $\mathcal{T}_j$ of $\mathcal{C}_j$, where $Q = \{0, 1, ..., v-1\}$ and $(Q, \circ)$ is an idempotent quasigroup of order v satisfying $s \circ t = \frac{s+t}{2}$ (for any $s, t \in Q$). Let $\alpha = (0, 1, ..., v-1)$ be a permutation of order v. Now we construct uv cyclic triple systems $\mathcal{A}_{ij}$

$(i \in I_u, j \in Q)$, which forms an $CLMTS(uv+2)$ on $X = \{a,b\} \bigcup (I_u \times Q)$, as follows:

(1) $\langle (x,s),(y,t),(z,(s \circ t)\alpha^j)\rangle$ provided $x,y,z \in I_u$, $\langle x,y,z\rangle \in \mathcal{B}_i$, and $s,t \in Q$;

(2) $\langle (x,s),(x,t),(y,(s \circ t)\alpha^j)\rangle$ provided $x,y \in I_u$, $\langle a,x,y\rangle \in \mathcal{B}_i$, $s \neq t \in Q$;

(3) $\langle a,(x,s),(y,s\alpha^j)\rangle$; $\langle b,(y,s\alpha^j),(x,s)\rangle$ provided $x,y \in I_u$, $\langle a,x,y\rangle \in \mathcal{B}_i$ and $s \in Q$;

(4) $\langle (i,s),(i,t),(i,w)\rangle$ provided $\langle s,t,w\rangle \in \mathcal{C}_j$, whenever a or b appears for s,t,w, omit the first coordinate i.

Proof. In Lemma 3.4.4 (C) we proved that $\{(X,\mathcal{A}_{ij}) : i \in I_u, j \in Q\}$ is an $QLMTS(uv+2)$. Let $\mathcal{J}_{ij}$ consist of two parts of cyclic triples as follows:

(A) $\{\langle (x,n+j),(y,n),(z,n+\frac{3}{2}j)\rangle : n \in Q, \langle x,y,z\rangle_* \in \mathcal{J}_i\}$,

(B) $\{\langle (i,s),(i,t),(i,w)\rangle : \langle s,t,w\rangle \in \mathcal{T}_j\}$,

where $\langle x,y,z\rangle_*$ means that the position of x,y,z is fixed in $\langle x,y,z\rangle$. Obviously $\mathcal{J}_{ij}$ is a divisible subset of $\mathcal{A}_{ij}$.

In fact, since $((n+j)\circ n)\alpha^j = \frac{2n+j}{2}\alpha^j = (n+\frac{3}{2}j)$, $\mathcal{J}_{ij}$ is contained in part (1) and part (4) of above construction. And $|\mathcal{J}_i| = \frac{u-1}{3}$ and $|\mathcal{T}_j| = \frac{v-1}{3}$ imply $|\mathcal{J}_{ij}| = v \cdot \frac{u-1}{3} + \frac{v-1}{3} = \frac{uv-1}{3}$, and any element of $(I_n \times Q)\backslash\{(i,j)\}$ appears in $\mathcal{J}_{ij}$. So, $\mathcal{J}_{ij}$ is a partition of $(I_n \times Q)\backslash\{(i,j)\}$. Next, we only to need show that any ordered pair of distinct elements in $I_n \times Q$ is contained in some cyclic triple of $\bigcup\limits_{i,j} \mathcal{J}_{ij}$, we exhaust all possibilities as follows:

(1) $P = ((i,s),(i,t)), s \neq t \in Q$ appears in (B) of $\mathcal{J}_{ij}$, if (s,t) is contained in some cyclic triple of $\mathcal{T}_j$.

(2) $P = ((x,s),(y,t)), x \neq y \in I_n$, $s,t \in Q$. If (x,y) is contained in some cyclic triple of $\mathcal{J}_i$, i.e., there exists $z \in I_n$ such that $\langle x,y,z\rangle_*$ or $\langle z,x,y\rangle_*$ or $\langle y,z,x\rangle_* \in \mathcal{J}_i$, then P appears in (A) of $\mathcal{J}_{ij}$, where $j \equiv s-t$ or $\frac{2}{3}(s-t)$ or $2(s-t) \pmod v$. □

Corollary 3.4.8. If $v = \prod\limits_i q_i$, where all q_i are distinct prime power and $q_i \equiv 1 \pmod 6$, then there exists an $CLMTS(v+2)$.

Proof. By Lemma 3.4.6 an $CSLMTS(q_i+2)$ exists for all q_i. Repeatly applying Lemma 3.4.7, the existence of $CLMTS(v+2)$ follows.

□

Using $CLMTS$ and $LD-$design, the following Theorem has been obtained in [KangChang 3].

Theorem 3.4.9. If an $LD(n)$ and an $CLMTS(n+2)$ both exist, then an $LMTS(3n+1)$ exists.

Construction. Let $I_n = \{0,1,...,n-1\}, a, b \notin I_n$ and $Z_3 = \{0,1,2\}$. Let $LD(n) = (I_n, \mathcal{L}^1, \mathcal{L}^2, (\mathcal{L}_x)_{x\in I_n})$ with $(x,x,x,0) \in \mathcal{L}^1 \bigcap \mathcal{L}^2$ for any $x \in I_n$, and $CLMTS(n+2) = \{(\{a,b\} \bigcup I_n, \mathcal{A}_x) : x \in I_n\}$ with $\langle a,b,x\rangle, \langle b,a,x\rangle \in \mathcal{A}_x$ and divisible subset $\mathcal{J}_x$ of $\mathcal{A}_x$.

Define n systems $\mathcal{B}_x (x \in I_n)$ of cyclic triples on $X = (I_n \times Z_3) \bigcup \{\infty\}$ as follows, where the element $(u,k) \in I_n \times Z_3$ is denoted by u_k:

(1) $\langle x_0, x_2, x_1\rangle, \langle\infty, x_0, x_1\rangle, \langle\infty, x_1, x_2\rangle, \langle\infty, x_2, x_0\rangle$;

(2) $\langle\alpha_0, \beta_1, \gamma_2\rangle, \langle\alpha_0, \gamma_2, \beta_1\rangle$ provided $(\alpha, \beta, \gamma) \in \mathcal{L}_x$;

(3) $\langle u_k, v_k, w_k\rangle$ provided $\langle u,v,w\rangle \in \mathcal{A}_x \backslash \mathcal{J}_x$; $\langle\infty, u_k, v_k\rangle, \langle\infty, v_k, w_k\rangle, \langle\infty, w_k, u_k\rangle$ provided $\langle u,v,w\rangle \in \mathcal{J}_x$. Where $u,v,w \in I_n$ and $k \in Z_3$;

(4) $\langle x_k, y_{k+1}, z_{k+1}\rangle$ provided $\langle a,y,z\rangle \in \mathcal{A}_x, y,z \in I_n$ and $k \in Z_3$;

(5) $\langle x_k, s_{k-1}, t_{k-1}\rangle$ provided $\langle b,s,t\rangle \in \mathcal{A}_x, s,t \in I_n$ and $k \in Z_3$.

Define $n-1$ systems $\mathcal{C}_{i1} (i \in I_n \backslash \{0\})$ of cyclic triples on X as follows:

(1) $\langle x_0, z_2, y_1\rangle, \langle\infty, x_0, y_1\rangle, \langle\infty, y_1, z_2\rangle, \langle\infty, z_2, x_0\rangle$ provided $(x,y,z,i) \in \mathcal{L}^1$;

$$(2)\quad \left.\begin{array}{lll} \langle x_0, u_1, v_1\rangle & \text{provided} & \langle a,u,v\rangle \in \mathcal{A}_y \\ \langle y_1, u_2, v_2\rangle & \text{provided} & \langle a,u,v\rangle \in \mathcal{A}_z \\ \langle z_2, u_0, v_0\rangle & \text{provided} & \langle a,u,v\rangle \in \mathcal{A}_x \end{array}\right\}, \text{ where } \langle x,y,z,i\rangle \in \mathcal{L}^1.$$

Define $n-1$ systems $\mathcal{C}_{i2} (i \in I_n \backslash \{0\})$ of cyclic triples on X as follows:

(1) $\langle x_0, y_1, z_2\rangle, \langle\infty, x_0, z_2\rangle, \langle\infty, z_2, y_1\rangle, \langle\infty, y_1, x_0\rangle$ provided $(x,y,z,i) \in \mathcal{L}^2$;

$$(2)\quad \left.\begin{array}{lll} \langle x_0, s_2, t_2\rangle & \text{provided} & \langle b,s,t\rangle \in \mathcal{A}_z \\ \langle y_1, s_0, t_0\rangle & \text{provided} & \langle b,s,t\rangle \in \mathcal{A}_x \\ \langle z_2, s_1, t_1\rangle & \text{provided} & \langle b,s,t\rangle \in \mathcal{A}_y \end{array}\right\}, \text{ where } \langle x,y,z,i\rangle \in \mathcal{L}^2.$$

Define a systme $\mathcal{B}_\infty$ of cyclic triples on X as follows:

(1) $\langle x_0, x_1, x_2\rangle$ provided $x \in I_n$;

(2) $\langle u_k, v_k, w_k\rangle$ provided $\langle u,v,w\rangle \in \bigcup_{x=0}^{n-1} \mathcal{J}_x,\ k \in Z_3$;

(3) $\langle\infty, x_0, x_2\rangle, \langle\infty, x_2, x_1\rangle, \langle\infty, x_1, x_0\rangle$ provided $x \in I_n$;

(4) $\langle x_0, y_1, z_2\rangle$ provided there exists $* \in I_n\backslash\{0\}$ such that $(x, y, z, *) \in \mathcal{L}^1$;

(5) $\langle x_0, z_2, y_1\rangle$ provided there exists $* \in I_n\backslash\{0\}$ such that $(x, y, z, *) \in \mathcal{L}^2$.

Then, $\{(X, \mathcal{B}_x) : x \in I_n \bigcup\{\infty\}\} \bigcup \{(X, \mathcal{C}_{ij}) : i \in I_n\backslash\{0\}, j = 1, 2\}$ is an $LMTS(3n + 1)$.

Proof. 1° Each $\mathcal{B}_x$ is an $MTS(3n + 1)$.

Direct calculation shows that $\mathcal{B}_x$ contains $\frac{3n(3n+1)}{3}$ cyclic triples, just the number we expected. We only need to show that every ordered pair P of distinct elements of the set $(I_n \times Z_3) \bigcup\{\infty\}$ is contained in some cyclic triple of $\mathcal{B}_x$. All the possibilities are exhausted as follows:

(1) $P = (x_i, x_j)(i \neq j \in Z_3)$ is contained in (1) of $\mathcal{B}_x$;

(2) $P = (x_i, u_i)$ (or $(u_i, x_i), u \in I_n\backslash\{x\}, i \in Z_3$) is contained in (3) of $\mathcal{B}_x$;

(3) $P = (x_i, u_j)(u \in I_n\backslash\{x\}, i \neq j \in Z_3)$ is contained in (4) or (5) of $\mathcal{B}_x$ if $j \equiv i + 1$ or $j \equiv i - 1 \pmod 3$. Similarly for $P = (u_j, x_i)$;

(4) $P = (u_i, v_j)(u, v \in I_n\backslash\{x\}, i \neq j \in Z_3)$ is contained in (2) of $\mathcal{B}_x$;

(5) $P = (u_i, v_i)(u \neq v \in I_n\backslash\{x\}, i \in Z_3)$. There exists $w \in \{a, b\} \bigcup I_n$ such that $\langle u, v, w\rangle \in \mathcal{A}_x$. If $w = a$ or b then P is contained in (4) or (5) of $\mathcal{B}_x$, otherwise in (3) of $\mathcal{B}_x$;

(6) $P = (\infty, u_k)(u \in I_n, k \in Z_3)$. If $u = x$. then P is contained in (1) of $\mathcal{B}_x$, otherwise in (3) of $\mathcal{B}_x$. Similarly for $P = (u_k, \infty)$.

2° Each $\mathcal{C}_{i1}(i \in I_n\backslash\{0\})$ is an $MTS(3n + 1)$.

$\mathcal{C}_{ij}$ contains $4n + 3n(n - 1) = n(3n + 1)$ cyclic triples. All ordered pairs P of the set $(I_n \times Z_3) \bigcup\{\infty\}$ occur as follows.

(1) $P = (u_k, v_k)(u \neq v \in I_n, k \in Z_3)$ is contained in (2) of $\mathcal{C}_{i1}$. For example, consider the case $k = 0$. There exists $x, z \in I_n$, such that $\langle a, u, v\rangle \in \mathcal{A}_x$ and $(x, *, z, i) \in \mathcal{L}^1$. Then, we have $\langle z_2, u_0, v_0\rangle \in \mathcal{C}_{i1}$. The remaining cases are similar.

(2) $P = (u_t, u_k)(u \in I_n, t \neq k \in Z_3)$ is contained in (2) of $\mathcal{C}_{i1}$. For example, consider the case $t = 0, k = 1$ (similar for other case of t and k). There exists $y \in I_n\backslash\{u\}$ such that $(u, y, *, i) \in \mathcal{L}^1$. Furthermore, there exists $v \in I_n\backslash\{u, y\}$ such that $\langle b, v, u\rangle \in \mathcal{A}_y$. Then, we have $\langle u_0, u_1, v_1\rangle \in$

C_{i1}.

(3) $P = (u_t, v_k)(u \neq v \in I_n, t \neq k \in Z_3)$ We only consider the case $t = 2, k = 1$ (other cases are similar). There exist $x, z \in I_n \backslash \{v\}$ such that $(x, v, z, i) \in \mathcal{L}^1$. If $z = u$, then P is contained in $\langle x_0, u_2, v_1 \rangle$ of part (1). If $z \neq u$, then there exists $w \in I_n \backslash \{z, u\}$ such that $\langle a, w, u \rangle \in \mathcal{A}_x$. Therefore, P is contained in $\langle v_1, w_2, u_2 \rangle$ of part (2) of C_{i1}.

(4) $P = (\infty, u_k)(u \in I_n, k \in Z_3)$. If $k = 1$ (other cases are similar). There exist $x, z \in I_n$ such that $(x, y, z, i) \in \mathcal{L}^1$. Then, P is contained in (1)of C_{i1}. Similarly for $P = (u_k, \infty)$.

3° Each $C_{i2}(i \in I_n \backslash \{0\})$ and $\mathcal{B}_\infty$ are all $MTS(3n + 1)$, too.

4° $\{\mathcal{B}_x : x \in I_n \bigcup \{\infty\}\} \bigcup \{C_{ij} : i \in I_n \backslash \{0\}, j = 1, 2\}$ is an $LMTS(3n+1)$.

We only need to show that every cyclic triple T from $(I_n \times Z_3) \bigcup \{\infty\}$ is contained in some $\mathcal{B}_x$ or C_{ij} above. All the possibilities are exhausted as follows:

(1) $T = \langle u_k, v_k, w_k \rangle (u \neq v \neq w \neq u \in I_n, k \in Z_3)$ is contained in (3) of $\mathcal{B}_x$ if $\langle u, v, w \rangle \in \mathcal{A}_x \backslash \mathcal{J}_x$. Otherwise in (2) of $\mathcal{B}_\infty$.

(2) $T = \langle u_0, v_1, w_2 \rangle$ or $\langle u_0, w_2, v_1 \rangle (u, v, w \in I_n)$. If $u = v = w$ then T is contained in (1) of $\mathcal{B}_u$ or (1) of $\mathcal{B}_\infty$. For other cases, by the properties of $LD(n)$, $\langle u, v, w \rangle \in \mathcal{L}_x$ or $(u, v, w, i) \in \mathcal{L}^j$, T is contained in (2) of $\mathcal{B}_x$ or in (1) of C_{ij} or in (4), (5) of $\mathcal{B}_\infty$.

(3) $T = \langle u_t, v_k, w_k \rangle (u, v, w \in I_n, t \neq k \in Z_3, v \neq w)$ leads to two cases.

Case 1. $k \equiv t + 1 \pmod 3$, as for exmple $t = 0, k = 1$. There exists $x \in I_n \backslash \{v, w\}$ such that $\langle a, v, w \rangle \in \mathcal{A}_x$. If $x = u$, then T is contained in (4) of $\mathcal{B}_x$, or else in (2) of C_{i1} provided $(u, x, *, i) \in \mathcal{L}^1$. The other cases are similar.

Case 2. $k \equiv t - 1 \pmod 3$, as for exmple $t = 1, k = 0$. There exists $x \in I_n \backslash \{v, w\}$ such that $\langle b, v, w \rangle \in \mathcal{A}_x$. If $x = u$, then T is contained in (5) of $\mathcal{B}_x$, else in (2) of C_{i2} provided $(x, u, *, i) \in \mathcal{L}^2$. The other cases are similar.

(4) $T = \langle \infty, x_k, x_{k+1} \rangle$ is contained in (1) of $\mathcal{B}_x$.

(5) $T = \langle \infty, x_k, x_{k-1} \rangle$ is contained in (3) of $\mathcal{B}_\infty$.

(6) $\langle \infty, u_k, v_k \rangle$ is contained in (3) of $\mathcal{B}_\infty$, where $\langle u, v, * \rangle \in \mathcal{J}_x$.

(7) $\langle \infty, u_k, v_{k+1} \rangle$ is contained in (1) of C_{i1}, where $\langle u, v, *, i \rangle$ or $(*, u, v, i)$ or $(v, u, *, i) \in \mathcal{L}^1$.

(8) $\langle \infty, u_k, v_{k-1} \rangle$ is contained in (1) of C_{i2}, where $\langle u, *, v, i \rangle$ or $(*, v, u, i)$ or $(v, u, *, i) \in \mathcal{L}^2$.

□

3.5 Spectrum of $LMTS(v, \lambda)$

It is well known that an $LMTS(v, \lambda)$ contains $\frac{v-2}{\lambda}$ $MTS(v, \lambda)$s. Thus, there exists an $LMTS(v, \lambda)$ only if

$$\lambda \leq v - 2, \; v \equiv 2 \pmod{\lambda}, \text{ if } \lambda \not\equiv 0 \pmod 3 \text{ then } v \not\equiv 2 \pmod 3 \qquad (*)$$
$$\text{and if } \lambda = 1 \text{ then } v \neq 6.$$

The existence spectrum of $LMTS(v, 1)$ has been completed in the last section. For any λ, the completion of the spectrum of $LMTS(v, \lambda)$ depends on determining the existence of $LMTS(v, 3)$.

3.5.1 $LMTS(v, 3)$ for $v = 5, 8$ and 14

Lemma 3.5.1. There exists an $LMTS(v, 3)$ for v=5 and 8.

Proof. It is easy to construct an $LMTS(5, 3)$. Since an $MTS(5, 3)$ contains 20 cyclic triples and an $LMTS(5, 3)$ consists of one $MTS(5, 3)$, we have

$$LMTS(5, 3) = \{(I_5, \; \mathcal{B})\},$$

where $\mathcal{B}$ contains just all cyclic triples from I_5.

Next, construct an $LMTS(8, 3) = \{(S, \; \mathcal{B}_i) : \; 0 \leq i \leq 1\}$ as follows: For any 3-subset $\{x, y, z\}$ of $S = \{1, 2, 3, 4, 5, 6, 7, 8\}$, $x < y < z$,

$$\text{if } x + y + z \text{ is even then } \langle x, y, z \rangle \in \mathcal{B}_0 \text{ and } \langle x, z, y \rangle \in \mathcal{B}_1;$$

$$\text{if } x + y + z \text{ is odd then } \langle x, y, z \rangle \in \mathcal{B}_1 \text{ and } \langle x, z, y \rangle \in \mathcal{B}_0.$$

In fact, for any $1 \le x < y \le 8$, we have the enumeration of the number of even integers and odd integers in the open interval $(0,x)$, or (x,y), or $(y,9)$:

Table 3.4

x	y	$1 \le z < x$		$x < z < y$		$y < z \le 8$	
		z even	z odd	z even	z odd	z even	z odd
even	even	$\frac{x-2}{2}$	$\frac{x}{2}$	$\frac{y-x-2}{2}$	$\frac{y-x}{2}$	$\frac{8-y}{2}$	$\frac{8-y}{2}$
odd	odd	$\frac{x-1}{2}$	$\frac{x-1}{2}$	$\frac{y-x}{2}$	$\frac{y-x-2}{2}$	$\frac{9-y}{2}$	$\frac{7-y}{2}$
even	odd	$\frac{x-2}{2}$	$\frac{x}{2}$	$\frac{y-x-1}{2}$	$\frac{y-x-1}{2}$	$\frac{9-y}{2}$	$\frac{7-y}{2}$
odd	even	$\frac{x-1}{2}$	$\frac{x-1}{2}$	$\frac{y-x-1}{2}$	$\frac{y-x-1}{2}$	$\frac{8-y}{2}$	$\frac{8-y}{2}$

By the table, it is not difficult to see that, for given $x < y$, the total number of z satisfying $x + y + z \equiv 0$ (1, resp.) mod 2 is just 3. Thus, each ordered pair $(x,\ y)$ (or $(y,\ x)$) appears just in three cyclic triples of $\mathcal{B}_i$ ($i = 0$ or 1). □

Lemma 3.5.2. There exists an $LMTS(14,3)$.

Construction. Let Z_3 and Z_4 be residue class rings modulo 3 and 4 respectively, $\infty_0,\infty_1 \notin Z_4 \times Z_3$. Construct four cyclic triple systems $\mathcal{B}_i$ ($i \in Z_4$) on the set $X = \{\infty_0, \infty_1\} \cup (Z_4 \times Z_3)$ as follows:

part 1. $\langle (i+1,x),(i+2,y),(i+3,z)\rangle$ and $\langle (i+3,x),(i+2,y),(i+1,z)\rangle$ where $x,y,z \in Z_3$ and $x+y+z=0$.

part 2. $\langle (i,x),(i+1,y),(i+2,z)\rangle$ and $\langle (i+3,x),(i+2,y),(i+1,z)\rangle$, where $x,y,z \in Z_3$ and $x+y+z=1$.

part 3. $\langle (i,x),(i+2,y),(i+3,z)\rangle$ and $\langle (i,x),(i+3,y),(i+1,z)\rangle$, where $x,y,z \in Z_3$ and $x+y+z=2$.

part 4. $\langle (i,x),(j,y),(j,z)\rangle$ where $j \in Z_4 \backslash \{i\}$, $x,\ y,\ z \in Z_3$ and $y \ne z$.

part 5. $\langle \infty_k,(i+1,x),(i+2,y)\rangle$, $\langle \infty_k,(i+2,x),(i+3,y)\rangle$ and $\langle \infty_k,(i+3,x),(i+1,y)\rangle$, where $x,y \in Z_3$ and $k = \frac{1-(-1)^i}{2}$.

part 6. $\langle \infty_k,(i+1,x),(i+3,y)\rangle$, $\langle \infty_k,(i+3,x),(i+2,y)\rangle$ and $\langle \infty_k,(i+2,x),(i+1,y)\rangle$, where $x,y \in Z_3$ and $k = \frac{1+(-1)^i}{2}$.

part 7. An $MTS(5,3)$ on the set $\{\infty_0,\ \infty_1\} \cup (\{i\} \times Z_3)$.

Proof. Each $\mathcal{B}_i$ contains

$$3 \times (2 \times 3^2) + 3 \times (2 \times 3^2) + 2 \times (3 \times 3^2) + 5 \times 4 = 182 = 14 \times 13$$

cyclic triples, just the number expected.

Obviously, each ordered pairs $(\infty_k,\ \infty_{1-k})$, $(\infty_k,\ (j,x))$ and $((j,x),\ \infty_k)$ are contained in just three blocks of part 5, 6 and 7, respectively. In every part, for given first coordinate pair $(j,k) \in Z_4 \times Z_4$, the corresponding second coordinate pairs run over $Z_3 \times Z_3$, and the repeated number of each one in $Z_3 \times Z_3$ is 1 (in part 1, 2, 3, 5 and 6) or 2 (in part 4 for (x,y) and (z,x)) or 3 (in part 4 for (y,z) and in part 7). For any pair of $Z_4 \times Z_4$, as the first coordinate pair, (i,i) appears in part 7, (j,j) appears in part 4, (i,j) appears in part 2 or 3 or 4, and (j,k) appears three times in part 1, 2, 3, 5, 6, where $j \neq k \in Z_4 \setminus \{i\}$.

In order to show that every cyclic triple T from X appears in some $\mathcal{B}_i$, we have

$T = \langle \infty_k,\ \infty_{1-k},\ (i,x) \rangle$ in part 7 of $\mathcal{B}_i$;

$T = \langle \infty_k,\ (i,x),\ (i,y) \rangle$ in part 7 of $\mathcal{B}_i$;

$T = \langle \infty_k,\ (i,x),\ (j,y) \rangle$ in part 6 or part 7 of some $\mathcal{B}_t$ (see the following table):

Table 3.5

$\mathcal{B}_t$	(i, j) in $\langle \infty_0, (i,x), (j,y) \rangle$	(i, j) in $\langle \infty_1, (i,x), (j,y) \rangle$
$\mathcal{B}_0$	(1,2), (2,3), (3,1)	(2,1), (3,2), (1,3)
$\mathcal{B}_1$	(3,2), (0,3), (2,0)	(2,3), (3,0), (0,2)
$\mathcal{B}_2$	(3,0), (0,1), (1,3)	(0,3), (1,0), (3,1)
$\mathcal{B}_3$	(1,0), (2,1), (0,2)	(0,1), (1,2), (2,0)

$T = \langle (i,x), (i,y), (i,z) \rangle$ in part 7 of $\mathcal{B}_i$;

$T = \langle (i,x), (j,y), (j,z) \rangle$ in part 4 of $\mathcal{B}_i$, where $j \neq i$;

$T = \langle (i,x), (j,y), (k,z) \rangle$ in part 1, 2, 3 of some $\mathcal{B}_t$. For this we only need to point out that the second coordinates in part 1, 2, 3 satisfy the equation $x + y + z$=0, 1, 2 on the set Z_3 respectively, and that the cyclic different triples of the first coordinates in part 1, 2, 3 are $\langle 1,1,2 \rangle$ and $\langle 3,3,2 \rangle$, which include all types of cyclic triples from Z_4. □

3.5.2 A recursive construction

Lemma 3.5.3. For $u \geq 3$ if there exists an $LMTS(u+2,1)$, then there exists an $LMTS(3u+2,3)$.

Construction. Let $(Q,\circ)$ be an idempotent quasigroup of order u, Z_3 be the residue class ring modulo 3, $\infty_1, \infty_2 \notin Z_3 \times Q$ and α be a cyclic permutation of length u over Q. Let

$$\{(\{\infty_1,\ \infty_2\} \cup Q,\ \mathcal{A}_r) :\ r \in Q\}$$

be an $LMTS(u+2,1)$. We construct u cyclic triple systems $\mathcal{B}_r$ $(r \in Q)$ on the set $X{=}\{\infty_1,\ \infty_2\}{\cup}(Z_3 \times Q)$ as follows:

part 1. $\langle (i,x),(i,y),(i,z)\rangle$, where $i \in Z_3$ and $\langle x,y,z\rangle \in \mathcal{A}_r$ (whenever ∞_1 or ∞_2 appears for x,y,z, omit the first coodinate i).

part 2. $\langle \infty_s,(i,x),(i\pm 1, x\alpha^r)\rangle$, where $i \in Z_3$, $x \in Q$ and $s=1,2$.

part 3. $\langle (i,x),(i,y),(i\pm 1,(x\circ y)\alpha^r)\rangle$, where $i \in Z_3$ and $x \neq y \in Q$.

part 4. $\langle (0,x),(1,y),(2,(x\circ y)\alpha^r)\rangle$ and $\langle (0,x),(2,y),(1,(x\circ y)\alpha^r)\rangle$, where $x,y \in Q$.

Then, $\{(X,\ \mathcal{B}_r) :\ r \in Q\}$ will be an $LMTS(3u+2,3)$.

Proof. Direct calculation shows that each $\mathcal{B}_r$ contains

$$(u+2)(u+1)+12u+6u(u-1)+2u^2 = (3u+2)(3u+1)$$

cyclic triples, just the number expected. Taking note of that $\{(x,y,(x\circ y)\alpha^r) :\ x,y \in Q\}$ is an orthogonal array $OA(n,3)$ and $(x\circ x)\alpha^r = x\alpha^r$, it is easy to verify that every ordered pair of distinct elements of X is contained in three cyclic triples of $\mathcal{B}_r$. Thus, each $\mathcal{B}_r$ is an $MTS(3u+2, 3)$.

In order to show that $\{(X,\ \mathcal{B}_r) :\ r \in Q\}$ is an $LMTS(3u+2,3)$, we only need to show that each cyclic triple T from X is contained in some $\mathcal{B}_r$ mentioned above. All the possibilities are exhausted as follows:

(i) $T = \langle \infty_s,\ \infty_{3-s},\ (i,x)\rangle$ is contained in part 1 of $\mathcal{B}_r$, if $\langle \infty_s,\ \infty_{3-s},\ x\rangle \in \mathcal{A}_r$.

(ii) $T = \langle \infty_s,\ (i,x),\ (i,y)\rangle$ is contained in part 1 of $\mathcal{B}_r$, if $\langle \infty_s,\ x,\ y\rangle \in \mathcal{A}_r$.

(iii) $T = \langle \infty_s,\ (i,x),\ (j,y)\rangle$, where $i \neq j$, is contained in part 2 of $\mathcal{B}_r$, if $y = x\alpha^r$.

(iv) $T = \langle (i,x),\ (i,y),\ (i,z)\rangle$ is contained in part 1 of $\mathcal{B}_r$, if $\langle x,\ y,\ z\rangle \in \mathcal{A}_r$.

(v) $T = \langle (i,x),\ (i,y),\ (j,z)\rangle$, where $i \neq j$, is contained in part 3 of $\mathcal{B}_r$, if $z = (x \circ y)\alpha^r$.

(vi) $T = \langle (0,x),\ (i,y),\ (j,z)\rangle$, where $i \neq j \in Z_3 \setminus \{0\}$, is contained in part 4 of $\mathcal{B}_r$, if $z = (x \circ y)\alpha^r$.

This completes the proof. □

Theorem 3.5.4. There exists an $LMTS(3u+2,3)$ for $u \not\equiv 0 \pmod 3$.

Proof. By Theorem 3.4.1, there exists an $LMTS(u+2,1)$ for $u \not\equiv 0 \pmod 3$ and $u \neq 4$. Thus, by Lemma 3.5.3, there exists an $LMTS(3u+2,3)$ for $u > 4$ and $u \not\equiv 0 \pmod 3$. But, by Lemmas 3.5.1 and 3.5.2, there exist $LMTS(3\times 1+2,\ 3)$, $LMTS(3\times 2+2,\ 3)$ and $LMTS(3\times 4+2,\ 3)$. Thus, there exists an $LMTS(3u+2,3)$ for any $u \not\equiv 0 \pmod 3$. □

Corollary 3.5.5. There exists an $LMTS(3\cdot 2^n+2,\ 3)$ for any non-negative integer n.

3.5.3 The spectrum of $LMTS(v,3)$

An $S(t,K,v)$, $t,\ v \in$N, $K \subset$N, is a pair $(X,\ \mathcal{B})$, where X is a v-set and $\mathcal{B}$ is a collection of subsets of X, called blocks, such that every t-subset of X is contained in exactly one block and such that $|B| \in K$ for all $B \in \mathcal{B}$. An $LS(\lambda,1;\ t,(k,K),v)$, $k \geq t$, $\lambda \in N \setminus \{0\}$ and $k \leq \min\{j;\ j \in K\}$, is a collection $\{(X,\mathcal{B}_r)\}_{r\in R}$ of $S(t,K,v) = (X,\mathcal{B}_r)$ such that $(X,\ \bigcup_{r\in R}\mathcal{B}_r)$ is an $S(k,K,v)$ and such that for each $B \in \bigcup_{r\in R}\mathcal{B}_r$ there are exactly $\lambda \binom{|B|-t}{k-t}$ elements r of R such that $B \in \mathcal{B}_r$ (Note that $\bigcup_{r\in R}\mathcal{B}_r$ denotes the ordinary union of the $\mathcal{B}_r$ and not the multiset-union). We write $LS(t,(k,K),v)$ instead of $LS(1,1;t,(k,K),v)$. It is well known that an $LS(\lambda, 1; t, (k, K), v)$, $v \geq$k, contains $\lambda\binom{v-t}{k-t}$ $S(t, K, v)$s.

An $S(2,(K_0, K_1, K_2), v)$ is a quadruple $(X, \infty_1, \infty_2, \mathcal{B})$, where $|X| = v-2$, $X \bigcap \{\infty_1, \infty_2\} = \phi$ and $\infty_1 \neq \infty_2$, such that $(X \bigcup \{\infty_1, \infty_2\}, \mathcal{B})$ is an $S(2, K_0 \bigcup K_1 \bigcup K_2, v)$ and such that $|B| \in K_i$ for all $B \in \mathcal{B}$ with $|B \bigcap \{\infty_1, \infty_2\}| = i$, $i \in \{0,1,2\}$. An $LS(\lambda,\ 1;\ 2,(3,\ K_0, K_1, K_2),\ v)$ will be a collection $(\{X, \infty_1, \infty_2, \mathcal{B}_r)\}_{r\in R}$ of $S(2,\ (K_0, K_1, K_2),\ v)$ such that $\{(X \bigcup \{\infty_1, \infty_2\},\ \mathcal{B}_r)\}_{r\in R}$ is an $LS(\lambda,\ 1;\ 2,\ (3,\ K_0 \bigcup K_1 \bigcup K_2),\ v)$. When λ=1, it is denoted by $LS(2,\ (3, K_0, K_1, K_2),\ v)$ briefly. If there exists an $LS(2,\ (3,\ K_0, K_1, K_2), v) = \{(X, \infty_1,\ \infty_2, \mathcal{B}_r)\ :\ r \in R\}$, define $\mathcal{B}_{(r,i)} = \mathcal{B}_r$, $i \in \{1,2\}$, we can get an $LS(2,1;2,(3,K_0,K_1,\ K_2),v) = \{(X, \infty_1, \infty_2,\ \mathcal{B}_{(r,i)});\ r \in R,\ i \in \{1,\ 2\}\}$. A *large set* of disjoint $GDD(t,k,ms;s\{m\})$, or $LGDD(t,k,ms;\ s\{m\})$, is a collection $\{(X, \mathcal{G}, \mathcal{A}_r)\}_{r\in R}$ of $GDD\ (t,k,ms;\ s\{m\})$, all having the same set of points and the same set of groups, such that, for every k-subset B of X satisfying $|B \bigcap G| \leq 1$ for all $G \in \mathcal{G}$, there is exactly one $r \in R$ with $B \in \mathcal{A}_r$. It is well known that there exists an $LGDD$(2,3,2s;$s\{2\}$) for any integer $s \equiv 0,\ 1 \pmod 3$ and $s \geq 3$ (see [Teir 8]). In particular, these exist for s=3 and 4.

Lemma 3.5.6. An $LS(2,(3,\{3\},\{3\},\{5\}), w+2)$, $w \geq 1$, exists if and only if $w \equiv 3 \pmod 6$.

Lemma 3.5.7. If an $LS(2,1;2,(3,K_0,\{3\},K_2),u+2)$ exists and if, for all $k \in K_0$, an $LGDD$(2, 3, 2k; $k\{2\}$) exists, then an $LS(2,(3,\{3,4\},\{3\},\ K_2'),2u+2)$ exists, where $K_2' = \{2k-2:\ k \in K_2\}$.

These results were obtained by Luc Teirlinck, and Lemma 3.5.6 and Lemma 3.5.7 above come from "construction 3.1" and "construction 4.1" in [Teir 8]. (Remark: In that paper, the "LS" in the above-mentioned conclusion is "GLS"—a kind of special "LS". Here the change doesn't matter for our research).

Lemma 3.5.8. For $w \equiv 3 \pmod 6$ and $n \geq 0$, there exists an $LS(2,(3,\{3,\ 4\},\{3\},\ \{3\cdot 2^n+2\}),\ 2^n w+2)$.

Proof. Use the method of induction on n. When $n = 0$, by Lemma 3.5.6, the conclusion holds. Suppose, for some n, there exists an $LS(2,(3,$

$\{3,4\}$, $\{3\}$, $\{3\cdot 2^n+2\}$), $2^n w+2$), then taking 2 copies of this large set we can get an $LS(2,1;\ 2,\ (3,\ \{3,4\},\ \{3\},\ \{3\cdot 2^n+2\}),\ 2^n w+2)$. And, since both $LGDD(2,3,6;3\{2\})$ and $LGDD(2,3,8;4\{2\})$ exist, there exists an $LS(2,(3,\ \{3,4\},\ \{3\},\ \{3\cdot 2^{n+1}+2\}),\ 2^{n+1}w+2)$ by Lemma 3.5.7, where $K_0=\{3,4\}$, $K_2=\{3\cdot 2^n+2\}$, $u=2^n w+2$ and $K_2'=\{2(3\cdot 2^n+2)-2\}$ $=\{3\cdot 2^{n+1}+2\}$. □

Theorem 3.5.9. There exists an $LMTS(v,3)$ if and only if $v\equiv 2 \pmod 3$ and $v\geq 5$.

Construction. Let $v=2^n w+2$, where n is non-negative integer and $w\equiv 3 \pmod 6$. By Lemma 3.5.8, there exists an $LS(2,(3,\{3,4\},\{3\},\{3\cdot 2^n+2\}),2^n w+2)=\{(X,\mathcal{B}_r^i)\ :\ r\in I_k,\ i\in I_3\}$, where X is a v-set, $I_3=\{0,\ 1,\ 2\}$, $I_k=\{0,\ 1,\cdots,k-1\}$ and $k=\frac{2^n w}{3}$.

For each $B\in \bigcup_{r\in I_k,i\in I_3}\mathcal{B}_r^i$, B appears in $|B|-2$ systems $\mathcal{B}_r^i$:

(1) If $|B|=3\cdot 2^n+2$, by the definition of $LS(2,(3,K_0,K_1,K_2),v)$, each $\mathcal{B}_r^i$ contains only one block with length $3\cdot 2^n+2$, which appears in $3\cdot 2^n$ systems $\mathcal{B}_r^i$. For the given $\{(X,\mathcal{B}_r^i)\ :\ r\in I_k,i\in I_3\}$ above, we can assign the label of these $\mathcal{B}_r^i$ such that the block with length $3\cdot 2^n+2$ in $\mathcal{B}_i^0$, $\mathcal{B}_i^1$, and $\mathcal{B}_i^2$ (for each $r\in I_k$) are the same. Thus, we can suppose $B\in \mathcal{B}_{r_k}^i$ ($i\in I_3$, $k=1,\ 2,\cdots,2^n$). By Corollary 3.5.5, there exists an $LMTS(3\cdot 2^n+2,\ 3)=\{(B,\ C_B(r_k))\ :\ 1\leq k\leq 2^n\}$.

(2) If $|B|=3$, denote $LMTS(3,1)=\{(B,\ D_B(s,t))\}$ when $B\in\mathcal{B}_s^t$.

(3) If $|B|=4$, suppose $B\in\mathcal{B}_{r_1}^{i_1}$ and $\mathcal{B}_{r_2}^{i_2}$, denote $LMTS(4,1)=\{(B,D_B(r_1,i_1)),\ (B,\ D_B(r_2,i_2))\}$.

Let A is the unique block with length $3\cdot 2^n+2$ in $\mathcal{B}_r^i$. Note that the block A in $\mathcal{B}_r^0$, $\mathcal{B}_r^1$ and $\mathcal{B}_r^2$ is same. Define

$$\mathcal{J}_r=\Big(\bigcup_{i=0}^{2}\{D_B(r,i):\ B\in\mathcal{B}_r^i,\ |B|=3,4\}\Big)\bigcup$$
$$\{C_A(r):\ A\in\mathcal{B}_r^0,\ |A|=3\cdot 2^n+2\},\ r\in I_k.$$

Then we get an expected $LMTS(2^n w+2,3)=\{(X,\ \mathcal{J}_r)\ :\ r\in I_k\}$.

Proof. An $LMTS(v,3)$ consists of $\frac{v-2}{3}$ disjoint systems $MTS(v,3)$, thus there exists an $LMTS(v,3)$ only if $v\equiv 2 \pmod 3$ and $v\geq 5$. As for the correctness of the construction above, we only need to point out that each $\mathcal{B}_r^i$ is a PBD on X with index 1, but each block A, $|A| = 3 \cdot 2^n + 2$, is replaced by an MTS with index 3 and each block B, $|B| = 3$ or 4, is replaced by an MTS with index 1. □

3.5.4 The existence of $LMTS(v, \lambda)$

Theorem 3.5.10. The existence spectrum of $LMTS(v, \lambda)$ is $v \equiv 2 \pmod{\lambda}$, $\lambda \leq v - 2$, if $\lambda \not\equiv 0 \pmod 3$, then $v \not\equiv 2 \pmod 3$ and if λ=1, then $v \neq 6$.

Proof. By the necessary conditions (∗) at the beginning of this section, we only need to proof the sufficiency.

For the case $\lambda \equiv 0 \pmod 3$, let $\lambda = 3t$, $t \geq 1$, X be a v-set and $v \equiv 2 \pmod{3t}$, $v \geq \lambda + 2$. By Theorem 3.5.9, since $v \equiv 2 \pmod 3$, there exists an $LMTS(v, 3) = \{(X, \mathcal{B}_i) : 1 \leq i \leq \frac{v-2}{3}\}$. Define

$$\mathcal{A}_k = \bigcup_{i=kt+1}^{(k+1)t} \mathcal{B}_i, \quad 0 \leq k \leq \frac{v-2}{3t} - 1,$$

then $\{(X, \mathcal{A}_k) : 0 \leq k \leq \frac{v-2}{3t} - 1\}$ is an $LMTS(v, 3t)$.

For the case $\lambda \not\equiv 0 \pmod 3$, let X be a v-set, $v \not\equiv 2 \pmod 3$, $v \equiv 2 \pmod{\lambda}$ and $v \geq \lambda + 2$. By Theorem 3.4.1, when $v \neq 6$, there exists an $LMTS(v, 1) = \{(X, \mathcal{B}_i) : 1 \leq i \leq v - 2\}$. Define

$$\mathcal{A}_k = \bigcup_{i=k\lambda+1}^{(k+1)\lambda} \mathcal{B}_i, \quad 0 \leq k \leq \frac{v-2}{\lambda} - 1,$$

then $\{(X, \mathcal{A}_k) : 0 \leq k \leq \frac{v-2}{\lambda} - 1\}$ is an $LMTS(v, \lambda)$.

Lastly, let us consider the case $v = 6$. By the necessary conditions (∗), the index λ only may take 2 or 4. It is trivial to construct an $LMTS(6, 4)$, which contains only one $MTS(6, 4)$. But an $LMTS(6, 2) = \{(S, \mathcal{B}_0), (S, \mathcal{B}_1)\}$ can be construct as follows: for any 3-subset $\{x, y, z\}$ of $S = \{1, 2, 3, 4, 5, 6\}$, $x < y < z$, if $x + y + z$ is even, then $\langle x, y, z \rangle \in \mathcal{B}_0$ and $\langle x, z, y \rangle \in \mathcal{B}_1$;

if $x + y + z$ is odd, then $\langle x, y, z \rangle \in \mathcal{B}_1$ and $\langle x, z, y \rangle \in \mathcal{B}_0$.

The correctness of the construction can be shown, similar to $LMTS(8, 3)$ in Lemma 3.5.1. □

Chapter 4

Large sets of Directed triple systems

4.1 Introduction and known results

Let X be a set of v elements ($v \geq 3$). A *transitive triple* from X is a collection of three ordered pairs $(x, y), (y, z), (x, z)$, where x, y, and z are distinct elements of X. We will denote the transitive triple $\{(x, y), (y, z), (x, z)\}$ by (x, y, z). A *directed triple system* on X is a pair $(X, \mathcal{B})$ where $\mathcal{B}$ is a collection of cyclic triples from X such that each ordered pair of distinct elements of X is covered by exactly λ transitive triples from $\mathcal{B}$. The number $| X |= v$ is called the *order* of the directed triple system $(X, \mathcal{B})$. For brevity, one denotes such system by $DTS(v, \lambda)$. It is easy to see that if $(X, \mathcal{B})$ is an $DTS(v, \lambda)$, then $| \mathcal{B} |= v(v-1)/3$. Thus, a necessary condition for the existence of a $DTS(v, \lambda)$ is $\lambda v(v-1) \equiv 0 \pmod 3$. When $\lambda = 1$, it becomes $v \equiv 0$ or $1 \pmod 3$. Usually, $DTS(v, 1)$ is briefly written by $DTS(v)$. In some papers, the system was once named as TTS (*transitive triple system*). But, we rather use the notation DTS to avoid confusing with twofold triple system (i.e. $TS(v, 2)$).

This kind of oriented design was studied first by S. H. Y. Hung and N. S. Mendelsohn [HungMen] in 1973. And, they proved that the existence spectrum for $DTS(v, 1)$ is just $v \equiv 0$ or $1 \pmod 3$. In 1980, J. R. Seberry and D. Skilicorn [SebSki] settled existence for all λ, that is

Theorem 4.1.1. A $DTS(v,\lambda)$ exists if and only if either $\lambda \equiv 1,2 \pmod 3$ and $v \equiv 0,1 \pmod 3$, or $\lambda \equiv 0 \pmod 3$ and $v \neq 2$.

As examples, we give $DTS(3)$ and $DTS(4)$. It is well known that there is only one nonisomorphic $DTS(3)$, i.e. $DTS(3) = (I_3, \mathcal{A})$, where $\mathcal{A} = \{(0,1,2),(2,1,0)\}$. And there are exactly three pairwise nonisomorphic $DTS(4) = (I_4, \mathcal{B}_i), i = 1,2,3$, whcih are listed as follows:

$\mathcal{B}_1$	$\mathcal{B}_2$	$\mathcal{B}_3$
(1, 2, 3)	(1, 2, 3)	(1, 2, 3)
(0, 3, 2)	(0, 3, 2)	(0, 3, 1)
(3, 0, 1)	(2, 0, 1)	(3, 0, 2)
(2, 1, 0)	(3, 1, 0)	(2, 1, 0)

It is easy to verify that there is no isomorphic mapping between $\mathcal{B}_i$ and $\mathcal{B}_j$ for any $i \neq j \in \{1,2,3\}$.

Let $C(X)$ be the set of all transitive triples from the set X of v elements. Then $| C(X) |= v(v-1)(v-2)$. The following problem arises quite naturally. Given a v-set X, $\lambda v(v-1) \equiv 0,1$ $(mod\ 3)$ and $v \geq 3$, is it always possible to partition $C(X)$ into $t = \frac{3(v-2)}{\lambda}$ subsets $\mathcal{B}_1, \mathcal{B}_2, \ldots, \mathcal{B}_t$ such that for $1 \leq i \leq t$ each $(X, \mathcal{B}_i)$ is a $DTS(v,\lambda)$? Such a collection of $DTS(v,\lambda)$'s is called a *large set* of pairwise disjoint $DTS(v,\lambda)$'s, we denote it by $LDTS(v,\lambda)$. By Theorem 4.1.1 and the definition of $LDTS$, it is easy to see that there exists an $LDTS(v,\lambda)$ only if

either $\lambda \not\equiv 0 \pmod 3$, $v \equiv 0,1 \pmod 3$ and $\lambda | 3(v-2)$,

or $\lambda \equiv 0 \pmod 3$, $\lambda | 3(v-2)$ and $v \neq 2$.

The research work on this problem started in 1983. C. C. Lindner and Anne Penfold Street presented some articles with respect to the existence of $LDTS(v,1)$. The author and Chang Yanxun, have thrown oneself into the work from 1990 to 1993 and have completely solved this problem for index $\lambda = 1$ and any λ, successively. After the completion of the spectrum of $LDTS(v,1)$, Luc Teirlinck raises a new way to solve the same problem based on the theory "Large sets with holes" (refer to [Teir 8]). Using his method, the existence spectrum for $LDTS(v,1)$ can be completed again, and that proof is extremely wonderful.

In this chapter, the process to complete the spectrum of $LDTS$ will be stated. In Section 4.2, the construction $LDTS(2^n+2)$, a key step to solve the case $\lambda = 1$, is given. In Section 4.3, the spectrum of $LDTS(v,1)$ will be completed, as well as some of other methods to construct $LDTS(v,1)$ will be introduced. Finally, in the Section 4.4, the spectrum of $LDTS(v,\lambda)$ will be obtained.

Up to 1990, all known results about $LDTS(v)$ (see [LinSt 1] and [Lin 2]) are listed as the following Lemmas.

Lemma 4.1.2. There exists an $LDTS(v)$ for $v \in \{3,4,6,7,18,24\}$.

Lemma 4.1.3. There exists an $LDTS(n+2)$ for $n \equiv 1,5 \pmod 6$.

Lemma 4.1.4. For $v \geq 3$ and $v \neq 6$, if there exists an $LDTS(v)$ then there exists an $LDTS(3v)$.

Lemma 4.1.5. For $v \geq 3$, if there exists an $LDTS(v+1)$ then there exists an $LDTS(3v+1)$.

Lemma 4.1.6. For $n \equiv 1,5 \pmod 6$, if there exists an $LDTS(v+2)$ then there exists an $LDTS(nv+2)$.

Below, our discussion will be started after these results. As examples, first let us give several $LDTS(v)$ for smaller orders $v = 3,4$ and 6.

Example 1. $LDTS(3) = \{(I_3, \mathcal{B}_i) : 1 \leq i \leq 3\}$:

$$\mathcal{B}_1 : (0,1,2),(2,1,0);$$
$$\mathcal{B}_2 : (0,2,1),(1,2,0);$$
$$\mathcal{B}_3 : (1,0,2),(2,0,1);$$

Example 2. $LDTS(4) = \{(I_4, \mathcal{B}_i) : 1 \leq i \leq 6\}$:

$\mathcal{B}_1$	$\mathcal{B}_2$	$\mathcal{B}_3$	$\mathcal{B}_4$	$\mathcal{B}_5$	$\mathcal{B}_6$
(0, 1, 2)	(0, 2, 3)	(0, 3, 1)	(2, 1, 0)	(3, 2, 0)	(1, 3, 0)
(1, 0, 3)	(2, 0, 1)	(3, 0, 2)	(3, 0, 1)	(1, 0, 2)	(2, 0, 3)
(2, 3, 0)	(3, 1, 0)	(1, 2, 0)	(0, 3, 2)	(0, 1, 3)	(0, 2, 1)
(3, 2, 1)	(1, 3, 2)	(2, 1, 3)	(1, 2, 3)	(2, 3, 1)	(3, 1, 2)

Example 3. $LDTS(6) = \{(Z_6, \mathcal{A}_i), (Z_6, \mathcal{B}_i) : i \in Z_6\}$. First, define

$$C = \{(0,4,5),(1,2,0),(2,3,4),(4,0,2),(5,3,0)\};$$
$$D = \{(0,3,1),(1,4,3),(2,1,5),(3,5,2),(5,4,1)\};$$
$$E = \{(3,1,0),(1,3,4),(1,2,5),(5,2,3),(4,5,1)\}.$$

Let $\mathcal{A} = C \bigcup D$ and $\mathcal{B} = C^{-1} \bigcup E$, where $C^{-1} = \{(x,y,z) : (z,y,x) \in C\}$.

Futheremore, let $\mathcal{A}_i = \mathcal{A} + i$ and $\mathcal{B}_i = \mathcal{B} + i$, $i \in Z_6$. This construction of $LDTS(6)$ is nonisomorphic to that one given in [LinSt 1].

4.2 $LDTS(2^n + 2, 1)$

4.2.1 A Symmetric $LDTS(2^4 + 2)$

In Chapter 3 (subsection 3.2.1) we introduced a kind of special large set of disjoint Mendelsohn triple systems ($LMTS$)—symmetric $LMTS$. Here, similarly, we define a kind of special $LDTS$—symmetric $LDTS$, $\{(X, \mathcal{B}_i)\}_i$, i.e., there exist $a \neq b \in X$ such that the following conditions hold for any i (where $x, y \notin \{a, b\}$):

(1) (a, b, x) (or (a, x, b) or (x, a, b)) $\in \mathcal{B}_i$ if and only if (x, b, a) (or (b, x, a) or (b, a, x)) $\in \mathcal{B}_i$;

(2) (a, x, y) (or (x, a, y) or (x, y, a)) $\in \mathcal{B}_i$ if and only if (b, y, x) (or (y, b, x) or (y, x, b)) $\in \mathcal{B}_i$.

Below, we will give a symmetric LDTS($2^4 + 2$), which is a key in this section and is taken from [Kang 2].

Let $F = GF(2^4)$, a finite field containing 2^4 elements, and g be a primitive element of F, $g^4 + g = 1$. Take two elements $\infty_1, \infty_2 \notin F$. We will construct

$$LDTS(18) = \{(\{\infty_1, \infty_2\} \bigcup F, \mathcal{B}_x^j) : x \in F, j \in \{1, 2, 3\}\}.$$

In each system $\mathcal{B}_x^j$ there are the following types of triples:

kernel-triple (i.e., triple containing ∞_1, ∞_2 and x),

∞_i–triple (i.e., triple containing ∞_i), i=1,2,

x–triple (i.e., triple containing x, but neither ∞_1 nor ∞_2),

y–triple (i.e., triple containing neither ∞_1, ∞_2 nor x).

For a given $x \in F$, each ordered pair (y, z), where $y \neq z \in F \backslash \{x\}$, can be uniquely written as $(y, \alpha x + (1 - \alpha) y)$ where $\alpha \in F \backslash \{0, 1\}$. We say that the ordered pair belongs to the pair class $\langle \alpha \rangle_x$ (briefly, PC $\langle \alpha \rangle_x$ or $\langle \alpha \rangle$). It is easy to see that each ordered pair of distinct elements of $F \backslash \{x\}$ belongs to a uniquely determined PC. In every y–triple there are three such ordered pairs, in every x–triple (or ∞_i–triple) there is only one such ordered pair, and there is none in every kernel-triple. Each PC contains $2^4-1=15$ pairwise distinct ordered pairs. The total number of the PCs (for given x) is $2^4-2=14$.

Each transitive triple (u,v,w), where $u\neq v\neq w\neq u\in F$, can be uniquely written as $(u,v,\lambda u+(1-\lambda)v)$, where $\lambda\in F\backslash\{0,1\}$. We say that the triple belongs to the triple class $[\lambda]$ (briefly TC $[\lambda]$). It is easy to see that each transitive triple (containing none of ∞_1 and ∞_2) belongs to a uniquely determined TC. Each TC contains $2^4(2^4-1)=240$ pairwise distinct triples. The total number of the TCs is $2^4-2=14$.

Define a mapping f_α (briefly f) from $F\backslash\{x\}$ into $F\backslash\{x\}$ as follows

$$f_\alpha: y\mapsto \alpha x+(1-\alpha)y,$$

where $x\in F$ is given and $\alpha\in F\backslash\{0,1\}$. We have $f^k(y)=(1-(1-\alpha)^k)x+(1-\alpha)^k y$, where $f^0(y)=y$ and $f^k(y)=f(f^{k-1}(y))$. So, $f^k(y)=y$ if and only if $(1-\alpha)^k=1$. Take note of $g^4+g=1$, g is a primitive element of F. We have the smallest period $k=15$ or 3, if $\alpha=g^4$ or g^5.

Lemma 4.2.1. There exists a partition of all ∞_1-triples and ∞_2-triples from the set $\{\infty_1,\infty_2\}\bigcup F$:

$$\bigcup_{j=1}^{3}\bigcup_{x\in F} C_x^j,$$

such that all ordered pairs of triples in each C_x^j are just

$$\{(\infty_i,u),(u,\infty_i):\ u\in F\backslash\{x\}, i=1,2\}\bigcup\{(y,z):$$

$$(y,z)\in\langle g^4\rangle_x \ \textit{or}\ \langle g^3\rangle_x\}.$$

Proof. By our statement mentioned above, we can denote all elements of $F\backslash\{x\}$ by $y_0,y_1,\ldots,y_{14}$ such that $y_{i+1}=f_{g^4}(y_i)$, i.e., $y_{i+1}=g^4x+gy_i$ (here and in what follows all subscripts are modulo 15). Let

$C_x^1=\{(\infty_1,y_{2i},y_{2i+1}),(y_0,\infty_1,y_1),(y_{2i-1},y_{2i},\infty_1),$
$(\infty_2,y_{2i+1},y_{2i}),(y_1,\infty_2,y_0),(y_{2i},y_{2i-1},\infty_2):\ 1\le i\le 7\};$

$C_x^2=\{(\infty_1,y_{2i+1},y_{2i+2}),(y_1,\infty_1,y_2),(y_{2i},y_{2i+1},\infty_1),$
$(\infty_2,y_{2i+2},y_{2i+1}),(y_2,\infty_2,y_1),(y_{2i+1},y_{2i},\infty_2):\ 1\le i\le 7\};$

$C_x^3 = \{(\infty_1, y_1, y_2), (y_i, \infty_1, y_{i+1}), (y_0, y_1, \infty_1),$
$(\infty_2, y_2, y_1), (y_{i+1}, \infty_2, y_1), (y_1, y_0, \infty_2) : 2 \le i \le 14\}.$

Obviously, $|C_x^1| = |C_x^2| = |C_x^3| = 30$. Note that $\{(y_i, y_{i+1}) \colon 0 \le i \le 14\} = \langle g^4 \rangle_x$ and $\{(y_{i+1}, y_i) \colon 0 \le i \le 14\} = \langle g^3 \rangle_x$. We can assert that the given construction satisfies all conditions of Lemma 4.2.1. □

Lemma 4.2.2. There exists a partition of all triples of the TC $[g^5]$ and $[g^{10}]$,

$$\bigcup_{j=1}^{3} \bigcup_{x \in F} E_x^j,$$

such that all ordered pairs of triples in each E_x^j are just

$$\{(y, z) : (y, z) \in \langle g^5 \rangle_x \text{ or } \langle g^{10} \rangle_x\}.$$

Proof. By remarks made before Lemma 4.2.1, we can partition all elements of $F \backslash \{x\}$ into circuits

$$\{y_{i,0}, y_{i,1}, y_{i,2}\}, \quad 1 \le i \le 5,$$

such that $y_{i,k+1} = f_{g^5}(y_{i,k})$, where $0 \le k \le 2$ and $y_{i,3} = y_{i,0}$. Let

$$E_x^1 = \{(y_{i,0}, y_{i,1}, y_{i,2}), (y_{i,2}, y_{i,1}, y_{i,0}) : 1 \le i \le 5\},$$

$$E_x^2 = \{(y_{i,1}, y_{i,2}, y_{i,0}), (y_{i,0}, y_{i,2}, y_{i,1}) : 1 \le i \le 5\},$$

$$E_x^3 = \{(y_{i,2}, y_{i,0}, y_{i,1}), (y_{i,1}, y_{i,0}, y_{i,2}) : 1 \le i \le 5\}.$$

Obviously, $|E_x^1| = |E_x^2| = |E_x^3| = 10$. Note that $g^5 + g^{10} = 1$ and

$$\{(y_{i,k}, y_{i,k+1}) : 1 \le i \le 5, 0 \le k \le 2\} = \langle g^5 \rangle_x,$$

$$\{(y_{i,k+1}, y_{i,k}) : 1 \le i \le 5, 0 \le k \le 2\} = \langle g^{10} \rangle_x.$$

Besides, we can verify that

$$\bigcup_{x \in F} \{(y_{i,k}, y_{i,k+1}, y_{i,k+2}) : 1 \le i \le 5, 0 \le k \le 2\} = [g^{10}],$$

$$\bigcup_{x \in F} \{(y_{i,k+2}, y_{i,k+1}, y_{i,k}) : 1 \le i \le 5, 0 \le k \le 2\} = [g^5].$$

Thus, the construction is true. □

Now, we can give a symmetric $LDTS(2^4+2) = \{((\{\infty_1,\infty_2\}\cup F, \mathcal{B}_x^j)\colon\ x\in F, j=1,2,3\}$ as follows:

$\mathcal{B}_x^1$:

(1)	$(\infty_1,\infty_2,x),(x,\infty_2,\infty_1)$		
(2)	C_x^1	$\langle 3\rangle\langle 4\rangle$	
(3)	E_x^1	$\langle 5\rangle\langle 10\rangle$	[5][10]
(4)	$(y,x,g^6x+g^{13}y)$	$\langle 6\rangle$	[13]
(5)	(y,g^8x+g^2y,g^7x+g^9y)	$\langle 8\rangle\langle 7\rangle\langle 9\rangle$	[3]
	$(y,g^{13}x+g^6y,g^{14}x+g^3y)$	$\langle 13\rangle\langle 14\rangle\langle 11\rangle$	[4]
	$(y,g^{12}x+g^{11}y,gx+g^4y)$	$\langle 12\rangle\langle 1\rangle\langle 2\rangle$	[1]

$\mathcal{B}_x^2$:

(1)	$(\infty_1,x,\infty_2),(\infty_2,x,\infty_1)$		
(2)	C_x^2	$\langle 3\rangle\langle 4\rangle$	
(3)	E_x^2	$\langle 5\rangle\langle 10\rangle$	[5][10]
(4)	(y,x,g^8x+g^2y)	$\langle 8\rangle$	[2]
(5)	$(y,g^2x+g^8y,g^{13}x+g^6y)$	$\langle 2\rangle\langle 13\rangle\langle 6\rangle$	[12]
	$(y,g^{14}x+g^3y,g^{11}x+g^{12}y)$	$\langle 14\rangle\langle 11\rangle\langle 7\rangle$	[11]
	$(y,g^9x+g^7y,g^{12}x+g^{11}y)$	$\langle 9\rangle\langle 12\rangle\langle 1\rangle$	[14]

$\mathcal{B}_x^3$:

(1)	$(x,\infty_1,\infty_2),(\infty_2,\infty_1,x)$		
(2)	C_x^3	$\langle 3\rangle\langle 4\rangle$	
(3)	E_x^3	$\langle 5\rangle\langle 10\rangle$	[5][10]
(4)	(y,x,g^2x+g^8y)	$\langle 2\rangle$	[8]
(5)	$(y,g^8x+g^2y,g^6x+g^{13}y)$	$\langle 8\rangle\langle 6\rangle\langle 12\rangle$	[6]
	$(y,g^{13}x+g^6y,g^7x+g^9y)$	$\langle 13\rangle\langle 7\rangle\langle 14\rangle$	[7]
	$(y,g^9x+g^7y,gx+g^4y)$	$\langle 9\rangle\langle 1\rangle\langle 11\rangle$	[9]

where $y\in F\backslash\{x\}$ and C_x^j, E_x^j are taken from Lemmas 4.2.1 and 4.2.2, respectively. What is written at the right side are the corresponding PC and TC. The PC $\langle\alpha\rangle_x$ and TC $[\alpha]$ are briefly denoted by $\langle m\rangle$ and $[m]$ if $\alpha=g^m$. For each element g^mx+g^ny, the equation $g^m+g^n=1$ holds. It is not difficult that the correctness of the construction has been verified (refer to [Kang 2]).

4.2.2 $16n+2$ construction

Let $F = GF(2^4)$, $Q = \{0, 1, ..., n-1\}, n \geq 3$ and $(Q, \circ)$ be an idempotent quasigroup of order n. On the set $F \times Q$ define a system of transitive triples $d_{y,z}(u,v) = \{((y,u),(y,v),(z,u\circ v)),((y,u),(z,u\circ v),(y,v)),((z,u\circ v),(y,u),(y,v))\}$, where $y \neq z \in F, u \neq v \in Q$. Then (refer to [LinSt 2])

$$D(y,z) = \bigcup_{u \neq v \in Q} d_{y,z}(u,v)$$

can be partitioned into three sets $D^1(y,z), D^2(y,z)$ and $D^3(y,z)$ such that

(1) The three triples in each $d_{y,z}(u,v)$ belong to different $D^j(y,z)$;

(2) If $u \neq v \in Q$, then each of the ordered pairs $((y,u),(z,v))$ and $((z,v),(y,u))$ belongs to exactly one triple in each of $D^1(y,z), D^2(y,z)$ and $D^3(y,z)$. Obviously, $|D^j(y,z)| = n(n-1), j = 1,2,3$.

Let $\infty_1, \infty_2 \notin F \times Q, X = \{\infty_1, \infty_2\} \bigcup (F \times Q)$ and α be a cycle of length n on Q. Let $\{(\{\infty_1, \infty_2\} \bigcup Q, \mathcal{L}_k^j) : k \in Q, j = 1,2,3\}$ be an $LDTS(n+2)$, and $\{(\{\infty_1, \infty_2\} \bigcup F, \mathcal{B}_x^j) : x \in F, j = 1,2,3\}$ be a symmetric $LDTS(2^4+2)$ given in the above subsection. Now, we can give the $16n+2$ construction,

$$LDTS(16n+2) = \{(X, \Omega_{x,k}^j) : x \in F, k \in Q, j = 1,2,3\}.$$

The system $\Omega_{x,k}^j$ consists of four parts of transitive triples as follows:

Part 1. $((x,u),(x,v),(x,w))$ provided $(u,v,w) \in \mathcal{L}_k^j$, whenever ∞_1 or ∞_2 appears for u, v, w omit the first coordinate x. This gives $(n+1)(n+2)/3$ triples.

Part 2. $((y,u),(z,v),(r,(u\circ v)\alpha^k))$ provided $(y,z,r) \in \mathcal{B}_x^j, y,z,r \in F$ and $u, v \in Q$. This gives $70n^2$ triples.

Part 3. The collection of triples obtained by replacing the unique pair (y,z) with $((y,u),(z,u\alpha^k))$ or $((y,u\alpha^k),(z,u))$ in each ∞_1-triple or ∞_2-triple of C_x^j respectively, where $u \in Q$. This gives $30n$ triples.

Part 4. The collection of triples obtained by replacing (z,w) with $(z, w\alpha^k)$ in each triple of $\{D^j(y,z) : (y,z) \in \langle g^4 \rangle_x\}$, where g is a primitive element of $F, g^4+g=1$, and $\langle g^4\rangle_x$ is a pair class. This gives $15n(n-1)$ triples.

Theorem 4.2.3. If there exists an $LDTS(n+2), n \geq 3$, then there exists an $LDTS(16n+2)$.

Proof. 1° Each $\Omega^j_{x,k}$ is a $DTS(16n+2)$. Direct calculation shows that each $\Omega^j_{x,k}$ contains

$$(n+1)(n+2)/3 + 70n^2 + 30n + 15n(n-1) = (16n+2)(16n+1)/3$$

transitive triples, just the number we expected. Therefore, we only need to show that every ordered pair P of distinct elements of the set X is contained in some triple of $\Omega^j_{x,k}$. All the possibilities are exhausted as follows ("P is contained by some triple in Part i" will be replaced briefly with "P in Part i"):

(1) $P = (\infty_i, \infty_{3-i}), (\infty_i, (x,u)), ((x,u), \infty_i)$, and $((x,u),(x,v))$ in Part 1, where $i=1,2, u \neq v \in Q$.

(2) $P = (\infty_i, (y,u))$ and $((y,u), \infty_i)$ in Part 3, where $i = 1,2, y \in F\backslash\{x\}$, and $u \in Q$.

(3) $P = ((y,u),(y,v)), y \in F\backslash\{x\}, u \neq v \in Q$. There exists $z \in F\backslash\{x\}$ such that $(y,z) \in \langle g^4\rangle_x$. So, P is covered by each triple of $d_{y,z}(u,v)$. Therefore, P in Part 4.

(4) $P = ((y,u),(z,v)), y \neq z \in F\backslash\{x\}, u,v \in Q$.

(i) If (y,z) or $(z,y) \in \langle g^4\rangle_x$, then there exists $w \in Q$ such that $(u \circ w)\alpha^k = v$. If $w=u$, i.e., $v = u\alpha^k$, then P is in Part 3. Or else $((y,u),(z,u\circ w))$ is covered by some triple of $D^j(y,z)$. Thus $P = ((y,u),(z,(u\circ w)\alpha^k))$ in Part 4.

(ii) If (y,z) and $(z,y) \notin \langle g^4\rangle_x$, then there exists $r \in F$ such that (y,z,r) or (y,r,z) or $(r,y,z) \in \mathcal{B}^j_x$. Thus $P \in ((y,u),(z,v),(r,w))$ or $((y,u),(r,w'),(z,v))$ or $((r,w''),(y,u),(z,v))$ in Part 2, where $w, w', w'' \in Q$ and $w = (u\circ v)\alpha^k$, $v = (u \circ w')\alpha^k, v = (w'' \circ u)\alpha^k$.

2° $\{X, \Omega^j_{x,k}) : x \in F, k \in Q, j = 1,2,3\}$ is an $LDTS(16n+2)$. We only need to show that every transitive triple T from X is contained in some $\Omega^j_{x,k}$ mentioned above. All the possibilities are exhausted as follows:

(1) $T = (\infty_i, \infty_{3-i}, (x,k)), (\infty_i, (x,k), \infty_{3-i})$ or $((x,k), \infty_i, \infty_{3-i}), i =$

$1,2,x \in F, k \in Q$. There exists $j \in \{1,2,3\}$ and $t \in Q$ such that T in Part 1 of $\Omega^j_{x,t}$.

(2) $T = (\infty_i, (y,u), (z,v))$ (or $((y,u), \infty_i, (z,v))$ or $((y,u,(z,v), \infty_i))$, $i = 1,2, (y,u) \neq (z,v) \in F \times Q$. If $y = z$ and (∞_i, u, v) (or (u, ∞_i, v) or $(u, v, \infty_i)) \in \mathcal{L}^j_k$, then T in Part 1 of $\Omega^j_{y,k}$. Or else, let $x = (z - gy)/g^4$ (if $i = 1$) or $x = (y - gz)/g^4$ (if $i = 2$). There exists $k \in Q$ such that $v = u\alpha^k$. Then T in part 3 of $\bigcup_{j=1}^3 \Omega^j_{x,k}$.

(3) $T = ((x,u),(x,v),(x,w)), x \in F, u \neq v \neq w \neq u \in Q$. There exist k and j such that $(u,v,w) \in \mathcal{L}^i_k$. Then T in Part 1 of $\Omega^j_{x,k}$.

(4) $T = ((y,u),(y,v),(z,w))$ (or $((y,u),(z,w),(y,v))$ or $((z,w),(y,u),$ $(y,v))), y \neq z \in F, u,v,w \in Q, u \neq v$. Let $x = (z - gy)/g^4$. There exists k such that $(u \circ v)\alpha^k = w$. Then T in Part 4 of $\bigcup_{j=1}^3 \Omega^j_{x,k}$.

(5) $T = ((y,u),(z,v),(r,w)), y \neq z \neq r \neq y \in F, u,v,w \in Q$. There exist x, j and k such that $(y,z,r) \in \mathcal{B}^j_x$ and $(u \circ v)\alpha^k = w$. Then T in Part 2 of $\Omega^j_{x,k}$.

This completes the proof. □

4.2.3 The conclusion

Theorem 4.2.4. For any integer $n \geq 1$, there exists an $LDTS(2^n+2)$.

Proof. First, by Lemma 4.1.2, we have $LDTS(4)$, $LDTS(6)$, and $LDTS$ (18). And $LDTS(4) \to LDTS(12)$, by Lemma 4.1.4. Furthermore, we have $LDTS(4) \to LDTS(10)$ and $LDTS(12) \to LDTS(34)$, by Lemma 4.1.5. So, for n =1, 2, 3, 4, 5, there exists an $LDTS(2^n+2)$. We use induction on n. Suppose there exists an $LDTS(2^k+2)$ for any $k \leq n (n \geq 5)$. By Theorem 4.2.3, since $2^{n-3} > 3$, there exists an $LDTS(16 \cdot 2^{n-3}+2)$=$LDTS(2^{n+1}+2)$. □

4.3 Spectrum of $LDTS(v,1)$

Now, by Section 4.2 and some known results stated in Section 4.1, we can give the existence theorem of $LDTS(v,1)$.

Theorem 4.3.1. For any $v \equiv 0$ or $1 \pmod 3$, there exists an $LDTS(v,1)$.

Proof. Let $v = 2^n u + 2$, where n is nonnegative integer and $u \equiv 1$ or $5 \pmod 6$. If $n = 0$, there exists an $LDTS(v,1)$ by Lemma 4.1.3. If $n \geq 1$, there exists an $LDTS(2^n + 2, 1)$ by Theorem 4.2.4; thereby there exists an $LDTS(v,1)$ by Lemma 4.1.6. □

In [Teir 8], Luc Teirlinck gave a new way to solve the existence spectrum of $LDTS(v,1)$. Even though the method has been discovered after the completion of the spectrum and the method need some further deep theory, it is worth to introduce his method here.

The Proposition 4.4 in [Teir 8] show that

Lemmma 4.3.2. An $LS(2,(3,\{3,4,6\}),v), v \geq 3$, exists if and only if $v \equiv 0$ or $1 (\mathrm{mod}\ 3)$ and $v \neq 7$.

Here, related concept and notation can be found in subsection 3.5.3. Taking 3 copies of an $LS(2,(3,\{3,4,6\}),v)$, we can obtain an $LS(3,1;2,(3,\{3,4,6\}),v)$ for $v \equiv 0, 1 (\mathrm{mod}\ 3)$ and $v \neq 7$. But,the contruction 2.6 in [Teir 8] give the existence of $LS(3,1;2,3,7)$. Therefore, we have

Lemma 4.3.3. An $LS(3,1;2,(3,\{3,4,6\}),v), v \geq 3$, exists if and only if $v \equiv 0$ or $1 (\mathrm{mod}\ 3)$.

Furthermore, Luc Teirlinck gave the following recursive construction (see Construction 7.1 of [Teir 8])

Lemma 4.3.4. If there exist both an $LS(3,1;2,(3,K),v), v \geq 3$, and an $LDTS(k)$ for all $k \in K$, then there exists an $LDTS(v)$.

Proof. Let $\{(X, \mathcal{B}_r) : r \in R\}$ be an $LS(3,1;2,(3,K),v)$. For each $B \in \bigcup_{r \in R} \mathcal{B}_r$, put $R_B = \{r \in R : B \in \mathcal{B}_r\}$ and let $\{(B, \mathcal{C}(B,r)) : r \in R_B\}$ be an $LDTS(|B|)$. Here, note that $|R_B| = 3(|B|-2)$ is just the number of $DTS(|B|)$ contained in an $LDTS(|B|)$. For each $r \in R$ define

$$\mathcal{A}_r = \bigcup_{B \in \mathcal{B}_r} \mathcal{C}(B,r).$$

Then $\{(X, \mathcal{A}_r) : r \in R\}$ is an $LDTS(v)$ expected. □

By Lemma 4.3.3 and Lemma 4.3.4, to solve the existence problem about $LDTS(v,1)$, we only need to construct $LDTS(k,1)$ for $k = 3, 4$, and 6. And these large sets for orders 3, 4 and 6 have been given in Section 4.1, as examples. Therefore, Theorem 4.3.1 can be obtained again.

Below, we give a directed method to construct $LDTS(2^{4m}+2)$ for any positive integer m, which is introduced from [ChangKang] and may present a way to construct nonisomophic $LDTS(2^n+2)$.

Let $F = GF(2^n)$ and $\infty_1, \infty_2 \notin F$. For a given $x \in F$, each ordered pair (y, z), where $y \neq z \in F \backslash \{x\}$, can be uniquely written as $(y, \alpha x + (1-\alpha)y)$ where $\alpha \in F\backslash\{0,1\}$. We say that the ordered pair belongs to the pair class $\langle\alpha\rangle_x$ (briefly $PC\langle\alpha\rangle_x$ or $\langle\alpha\rangle$). It is easy to see that each ordered pair of distinct elements of $F\backslash\{x\}$ belongs to a uniquely determined PC. Each PC contains 2^n-1 pairwise distinct ordered pairs. The total number of the PCs (for given x) is 2^n-2.

Each transitive triple (u, v, w), where u, v, w are distinct elements of F, can be uniquely written as $(u, v, \lambda u + (1-\lambda)v)$, where $\lambda \in F\backslash\{0,1\}$. We say that the triple belongs to the triple class $[\lambda]$ (briefly $TC[\lambda]$). It is easy to see that each transitive triple belongs to an uniquely determined TC. Each TC contains $2^n(2^n-1)$ pairwise distinct transitive triples. The total number of the TCs is 2^n-2.

For $\lambda \in F\backslash\{0,1\}$, the subset $\{\lambda, 1-\lambda, 1/\lambda, (1-\lambda)/\lambda, \lambda/(1-\lambda), 1/(1-\lambda)\} \subset F\backslash\{0,1\}$ is called a *sextuple* and denoted by $\{\lambda\}$ (or $\{1-\lambda\}$, $\{1/\lambda\}$, ..., since the sextuple is invariant for the distinct selection of the representative). It is easy to see that the set $F\backslash\{0,1\}$ can be partitioned into $(2^n-2)/6$ normal sextuples (which contains six distinct elements) if n odd, or into $(2^n-4)/6$ normal sextuples and one small sextuple $\{\varepsilon, \varepsilon, \varepsilon, \varepsilon^2, \varepsilon^2, \varepsilon^2\}$ if n even, where ε is a cubic primitive root. For convenience, the $TC[\varepsilon]$ and $[\varepsilon^2]$ are also called small TC.

It is well known that the field $F = GF(2^n)$ can be regarded as a subfield of the field $\overline{F} = GF(2^{mn})$, where m, n are positive integers. For even n, when all elements of $\overline{F}\backslash\{0,1\}$ are partitioned into sextuples, all elements of $\overline{F}\backslash F$ are just partitioned into some normal sextuples, because $\{\lambda\} = \{\lambda, 1-\lambda, 1/\lambda, (1-\lambda)/\lambda, \lambda/(1-\lambda), 1/(1-\lambda)\} \subset F\backslash\{0,1\}$ if $\lambda \in F\backslash\{0,1\}$ and the unique small sextuple of $\overline{F}\backslash\{0,1\}$ is in $F\backslash\{0,1\}$. Let S be the set of the representatives of normal sextuple in $\overline{F}\backslash F$.

For given $x \in \overline{F}$ and $\lambda \in S$ let

$$D_x^1(\lambda) = \{((1-\lambda)x+\lambda y, \lambda x+(1-\lambda)y, y), (y, \lambda x+(1-\lambda)y, (1-\lambda)x+\lambda y) : y \in \overline{F}\backslash\{x\}\};$$

$$D_x^2(\lambda) = \{(\lambda x+(1-\lambda)y, (1-\lambda)x+\lambda y, y), (y, (1-\lambda)x+\lambda y, \lambda x+(1-\lambda)y) : y \in \overline{F}\backslash\{x\}\};$$

$$D_x^3(\lambda) = \{((1-\lambda)x+\lambda y, y, \lambda x+(1-\lambda)y), (\lambda x+(1-\lambda)y, y, (1-\lambda)x+\lambda y) : y \in \overline{F}\backslash\{x\}\}.$$

Consider the mapping f_α (briefly f) from $F\backslash\{x\}$ onto $F\backslash\{x\}$

$$f_\alpha : y \to \alpha x + (1-\alpha)y.$$

Under the action of f, all elements of $F\backslash\{x\}$ are partitioned into $(2^n-1)/k$ circuits with odd length $k > 1$. Suppose these circuits are

$$M_i = (y_0^{(i)}, y_1^{(i)}, \ldots, y_{k-1}^{(i)}),$$

where $i = 1, 2, \ldots, (2^n-1)/k$ and $y_{t+1}^{(i)} = f(y_t^{(i)})$ $(0 \le t \le k-1$, here and in what follows, all subscripts are modulo $k)$. Let

$$C_x^1(\alpha, j) = \{(\infty_j, y_{2t}^{(i)}, y_{2t+1}^{(i)}), (y_0^{(i)}, \infty_j, y_1^{(i)}), (y_{2t-1}^{(i)}, y_{2t}^{(i)}, \infty_j) : 1 \le i \le (2^n-1)/k, 1 \le t \le (k-1)/2\};$$

$$C_x^2(\alpha, j) = \{(\infty_j, y_{2t+1}^{(i)}, y_{2t+2}^{(i)}), (y_1^{(i)}, \infty_j, y_2^{(i)}), (y_{2t}^{(i)}, y_{2t+1}^{(i)}, \infty_j) : 1 \le i \le (2^n-1)/k, 1 \le t \le (k-1)/2\};$$

$$C_x^3(\alpha, j) = \{(\infty_j, y_1^{(i)}, y_2^{(i)}), (y_t^{(i)}, \infty_j, y_{t+1}^{(i)}), (y_0^{(i)}, y_1^{(i)}, \infty_j) : 1 \le i \le (2^n-1)/k, 2 \le t \le k-1\}.$$

Now, take $n = 4$. Let $F = GF(2^4)$ be a subfield of $F = GF(2^{4m}), m \ge 1, X = \{\infty_1, \infty_2\} \bigcup F$ and g be a primitive element of F, where $g^4 + g = 1$. Define

A_x^1 $(x \in F)$:

$(1)_1$	$(\infty_1, \infty_2, x), (x, \infty_2, \infty_1)$		
$(2)_1$	$C_x^1(g^4, 1)$	$\langle 4\rangle$	
	$C_x^1(g^8, 2)$	$\langle 8\rangle$	
$(3)_1$	$(y, x, g^7x + g^9y)$	$\langle 7\rangle$	[9]
$(4)_1$	$(y, gx + g^4y, g^9x + g^7y)$	$\langle 1\rangle\langle 9\rangle\langle 14\rangle$	[2]
	$(y, g^2x + g^8y, g^{13}x + g^6y)$	$\langle 2\rangle\langle 13\rangle\langle 6\rangle$	[12]
	$(y, g^{11}x + g^{12}y, g^{12}x + g^{11}y)$	$\langle 11\rangle\langle 12\rangle\langle 3\rangle$	[4]
$(5)_1$	E_x^1	$\langle 5\rangle\langle 10\rangle$	[5][10]

$\mathcal{A}_x^2$ $(x \in F)$:

$(1)_2$	$(\infty_1, x, \infty_2), (\infty_2, x, \infty_1)$		
$(2)_2$	$C_x^2(g^4, 1)$	$\langle 4\rangle$	
	$C_x^2(g^8, 2)$	$\langle 8\rangle$	
$(3)_2$	$(y, x, g^{12}x + g^{11}y)$	$\langle 12\rangle$	$[11]$
$(4)_2$	$(y, g^{11}x + g^{12}y, g^{14}x + g^3y)$	$\langle 11\rangle\langle 14\rangle\langle 13\rangle$	$[14]$
	$(y, g^7x + g^9y, g^6x + g^{13}y)$	$\langle 7\rangle\langle 6\rangle\langle 1\rangle$	$[3]$
	$(y, g^3x + g^{14}y, g^9x + g^7y)$	$\langle 3\rangle\langle 9\rangle\langle 2\rangle$	$[13]$
$(5)_2$	E_x^2	$\langle 5\rangle\langle 10\rangle$	$[5][10]$

$\mathcal{A}_x^3$ $(x \in F)$:

$(1)_3$	$(x, \infty_1, \infty_2), (\infty_2, \infty_1, x)$		
$(2)_3$	$C_x^3(g^4, 1)$	$\langle 4\rangle$	
	$C_x^3(g^8, 2)$	$\langle 8\rangle$	
$(3)_3$	$(y, x, g^{13}x + g^6y)$	$\langle 13\rangle$	$[6]$
$(4)_3$	$(y, g^{14}x + g^3y, g^3x + g^{14}y)$	$\langle 14\rangle\langle 3\rangle\langle 12\rangle$	$[1]$
	$(y, g^2x + g^8y, g^{11}x + g^{12}y)$	$\langle 2\rangle\langle 11\rangle\langle 1\rangle$	$[7]$
	$(y, g^7x + g^9y, g^9x + g^7y)$	$\langle 7\rangle\langle 9\rangle\langle 6\rangle$	$[8]$
$(5)_3$	E_x^3	$\langle 5\rangle\langle 10\rangle$	$[5][10]$

where $y \in F\backslash\{x\}$, $C_x^i(a, j)$ are taken from the above-mentioned subconstruction, and E_x^i are taken from after Lemma 4.2.2. What is written at the right side are the corresponding PC and TC. The PC $\langle\alpha\rangle_x$ and TC $[\alpha]$ are briefly denoted by $\langle m\rangle$ and $[m]$ if $\alpha = g^m$. For each element $g^m x + g^n y$, the equation $g^m + g^n = 1$ holds. It is not hard to verify that $\{(X, \mathcal{A}_x^i)\colon\ x \in F, i = 1, 2, 3\}$ is an $LDTS(2^4 + 2)$ (refer to [Kang 2]).

Let $D_x^1 = \bigcup_{\lambda \in S} D_x^1(\lambda)$, $D_x^2 = \bigcup_{\lambda \in S} D_x^2(\lambda)$ and $D_x^3 = \bigcup_{\lambda \in S} D_x^3(\lambda)$, where $x \in F$ and $D_x^1(\lambda)$, $D_x^2(\lambda)$, $D_x^3(\lambda)$ are taken from the above-mentioned construction. Substituting the range "$y \in \overline{F}\backslash\{x\}$" for "$y \in F\backslash\{x\}$" in the part $(3)_1, (3)_2, (3)_3$ and $(4)_1, (4)_2, (4)_3$ of $\mathcal{A}_x^1, \mathcal{A}_x^2$ and $\mathcal{A}_x^3$ (also to the C_x^1, C_x^2, C_x^3 and E_x^1, E_x^2, E_x^3 in the part $(2)_1, (2)_2, (2)_3$ and $(5)_1, (5)_2, (5)_3$), we get $\mathcal{B}_x^1, \mathcal{B}_x^2$ and $\mathcal{B}_x^3$. Finally, we can obtain

$$LDTS(2^{4m} + 2) = \{(X, \mathcal{B}_x^i) : x \in \overline{F}, i = 1, 2, 3\},$$

where $\mathcal{B}_x^i = \mathcal{A}_x^i \bigcup \mathcal{D}_x^i$, $i = 1, 2, 3$.

By the method mentioned above, if there exists an $LDTS(2^t + 2)$ having similar structure to $LDTS(2^4+2)$ above, then an $LDTS(2^{tm}+2)$ can be obtained also.

4.4 Spectrum of $LDTS(v, \lambda)$

4.4.1 The existence of $LDTS(v, 3)$

The solution of $LDTS(v, \lambda)$ for larger λ requires only a complete solution for $LDTS(v, 3)$. In this section we will discuss the existence of $LDTS(v, 3)$. First, a recurrence construction on $LDTS(v, 3)$ is given as follows.

Theorem 4.4.1. If there exists an $LDTS(m + 2, 3)$ and $n \equiv 1, 5 \pmod 6$, then there exists an $LDTS(mn + 2, 3)$.

Construction. Let $Z_n = \{0, 1, ..., n-1\}, a \neq b, a, b \notin Z_n$ and $\{(\{a, b\} \bigcup I_m, \mathcal{D}_i)\colon\ i \in I_m\}$ be a $LDTS(m + 2, 3)$. And let $Q = (I_m, \circ)$ be an idempotent quasigroup of order m and $\alpha = (0, 1, ..., m-1)$ be a cycle of order m. Construct mn transitive triple systems $\mathcal{B}_{ij} (j \in Z_n, j \in I_m)$ on the set $X = \{a, b\} \bigcup (Z_n \times I_m)$ as follows.

(1) $((x, u), (y, v), (z, (u \circ v)\alpha^j))$ with $u, v \in I_v$, $x+y+z = 3i$ and x, y, z are distinct elements of Z_n. This gives $(n-1)(n-2)m^2$ transitive triples.

(2) $((x, u), (x, v), (y, (u \circ v)\alpha^j))$, $((x, u), (y, (u \circ v)\alpha^j), (x, v))$, $(y, (u \circ v)\alpha^j), (x, u), (x, v))$ with $x \neq y \in Z_n, u \neq v \in I_m$ and $x + 2y = 3i$. This gives $3m(m-1)(n-1)$ transitive triples.

(3) $(a, (x, u), (y, u\alpha^j))$, $((x, u), a, (y, u\alpha^j))$, $((x, u), (y, u\alpha^j), a)$, $(b, (y, u\alpha^j), (x, u))$, $((y, u\alpha^j), b, (x, u))$, $((y, u\alpha^j), (x, u), b)$, where $x \neq y \in Z_n$, $u \neq v \in I_m$ and $x + 2y = 3i$. This gives $6m(n-1)$ transitive triples.

(4) $((i, u), (i, v), (i, w))$ with $(u, v, w) \in \mathcal{D}_j$ (whenever a or b appears for u, v, w, omit the first coordinate i). This gives $(m+2)(m+1)$ transitive triples.

Proof. Direct calculation shows that each $\mathcal{B}_{ij}$ contains $(mn+2)(mn+1)$ transitive triples, just the number expected. It is easy to verify that every ordered pair P of distinct elements of X is contained in three transitive triples of $\mathcal{B}_{ij}$. To prove $\{(X, \mathcal{B}_{ij}) :\ i \in Z_n, j \in I_m\}$ is an $LDTS(v, 3)$, we only need to show that every transitive triple T of X is contained in some $\mathcal{B}_{ij}$. All the possibilities are exhausted as follows:

(i) $T = (a, (x, u), (y, v))(x \neq y \in Z_n, u, v \in I_m)$. There exist $i \in Z_n$ and $j \in I_m$ such that $x + 2y = 3i$ and $v = u\alpha^j$. Then T is contained in (3) of $\mathcal{B}_{ij}$ (It is similar for $((x, u), a, (y, v))$, $((x, u), (y, v), a)$, $(b, (x, u), (y, v))$, $((x, u), b, (y, v))$ and $((x, u), (y, v), b)$.

(ii) $T = ((i, u), (i, v), (i, w)), i \in Z_n, u, v, w$ are distinct elements of $\{a, b\} \bigcup I_m$ (whenever a or b appears for u, v, w, omit the first coordinate). There exists $j \in I_m$ such that $(u, v, w) \in \mathcal{D}_j$. Then T is contained in (4) of $\mathcal{B}_{ij}$.

(iii) $T = ((x, u), (x, v), (y, w))(x \neq y \in Z_n, u, v, w \in I_m$ and $u \neq v)$. There exists $i \in Z_n$, $j \in I_m$ such that $x + 2y = 3i$ and $w = (u \circ v)\alpha^j$, then T is contained in (2) of $\mathcal{B}_{ij}$ (It is similar for $T = ((x, u), (y, w), (x, v))$ and $((y, w), (x, u), (x, v))$.

(iv) $T = ((x, u), (y, v), (z, w))(x \neq y \neq z \neq x \in Z_n$ and $u, v, w \in I_m)$. There exist $i \in Z_n$ and $j \in I_m$ such that $x + y + z = 3i$ and $w = (u \circ v)\alpha^j$. Then T is contained in (1) of $\mathcal{B}_{ij}$.

This completes the proof. □

Two idempotent Latin squares are called *disjoint* if they do not agree in an off-diagonal position. A *large set* of idempotent Latin squares of order v is a collection of $v-2$ pairwise disjoint idempotent Latin squares of order v.

Lemma 4.4.2. [TeirLin] If $v \neq 2, 6, 14, 62$, then there exists a large set of idempotent Latin squares of order v.

Theorem 4.4.3. If there exists a large set of idempotent Latin squares of order v, then there exists an $LDTS(v, 3)$.

Proof. Let $X = \{1, 2, ..., v\}$ and let

$$A_k = (a_{ij}^{(k)})_{i,j=1}^{v}, \quad 1 \leq k \leq v - 2,$$

be a large set of idempotent Latin squares of order v.

Now we define $v - 2$ transitive triple systems as follows:

$$\mathcal{D}_k = \{(i, j, a_{ij}^{(k)}) : \ a_{ij}^{(k)} \in A_k \text{ and } i \neq j\}, \ k = 1, ..., v - 2.$$

Since each A_k is an idempotent Latin square, thus it is not difficult to see that each $(X, \mathcal{D}_k)$ is a $DTS(v,3)$, for $k = 1,2,...,v-2$. Futhermore, since A_i and A_j are disjoint for any $i \neq j$, it follows $(X, \mathcal{D}_i)$ and $(X, \mathcal{D}_j)$ are disjoint. Therefore, $\{(X, \mathcal{D}_k) : k = 1,2,...,v-2\}$ is an expected $LDTS(v,3)$. □

Lemma 4.4.4. There exists an $LDTS(14,3)$.

Construction. By Lemma 4.1.2, there exists an $LDTS(6,1) = \{(I_4 \bigcup\{a, b\}, \mathcal{B}_{ij})$: $i \in Z_3, j \in I_4\}$. And let $Q = (I_4, \circ)$ be an idempotent quasigroup of order 4, $\sigma = (0,1,2,3)$ be a cycle of order 4 on I_4. Now we can construct 12 transitive triple systems $\mathcal{D}_{ij}$ $(i \in Z_3, j \in I_4)$ on $X = (Z_3 \times I_4)\bigcup\{a,b\}$ as follows:

Each $\mathcal{D}_{0j}$ contains the following transitive triples:

(1) $((i,x),(i,y),(i,z))$, where $i \in Z_3, (x,y,z) \in \mathcal{B}_{0j}$ (whenever a or b appears for x,y,z, omit the first coordinate i). This part gives 30 transitive triples.

(2) $((0,u),(1,v),(2,(u\circ v)\alpha^j)), ((2,(u\circ v)\alpha^j),(1,v),(0,u))$, where $u,v \in I_4$. This part gives 32 transitive triples.

(3) $((i,u),(i,v),(i+1,(u\circ v)\alpha^j)), ((i+1,(u\circ v)\alpha^j),(i,v),(i,u))$, where $i \in Z_3, u \neq v \in I_4$. This part gives 72 transitive triples.

(4) $(a,(i,u),(i+1,u\alpha^j)), ((i+1,u\alpha^j),(i,u),a), (b,(i,u),(i+1,u\alpha^j)), ((i+1,u\alpha^j),(i,u),b)$, where $i \in Z_3$ and $u \in I_4$. This part gives 48 transitive triples.

Each $\mathcal{D}_{1j}$ contains the following transitive triples:

(1) $((i,x),(i,y),(i,z))$, where $i \in Z_3, (x,y,z) \in \mathcal{B}_{1j}$ (whenever a or b appears for x,y,z, omit the first coordinate i). This part gives 30 transitive triples.

(2) $((1,u),(2,v),(0,(u\circ v)\alpha^j)), ((0,(u\circ v)\alpha^j),(2,v),(1,u))$, where $u,v \in I_4$. This part gives 32 transitive triples.

(3) $((i,u),(i+1,(u\circ v)\alpha^j),(i,v)), ((i,v),(i-1,(u\circ v)\alpha^j),(i,u))$, where $i \in Z_3, u \neq v \in I_4$. This part gives 72 transitive triples.

(4) $((i,u),a,(i+1,u\alpha^j)), ((i+1,u\alpha^j),a,(i,u)), ((i,u),b,(i-1,u\alpha^j)), ((i-1,u\alpha^j),b,(i,u))$, where $i \in Z_3$ and $u \in I_4$. This part gives 48 transitive triples.

Each $\mathcal{D}_{2j}$ contains the following transitive triples:

(1) $((i,x),(i,y),(i,z))$, where $i \in Z_3, (x,y,z) \in \mathcal{B}_{2j}$ (whenever a or b appears for x,y,z, omit the first coordinate i). This part gives 30 transitive triples.

(2) $((2,u),(0,v),(1,(u \circ v)\alpha^j)), ((1,(u \circ v)\alpha^j),(0,v),(2,u))$, where $u,v \in I_4$. This part gives 32 transitive triples.

(3) $((i-1,(u \circ v)\alpha^j),(i,u),(i,v)), ((i,v),(i,u),(i-1,(u \circ v)\alpha^j))$, where $i \in Z_3, u \neq v \in I_4$. This part gives 72 transitive triples.

(4) $(a,(i,u),(i-1,u\alpha^j)), ((i-1,u\alpha^j),(i,u),a), (b,(i,u),(i-1,u\alpha^j)), ((i-1,u\alpha^j),(i,u),b)$, where $i \in Z_3$ and $u \in I_4$. This part gives 48 transitive triples.

Proof. 1°. Each of $(X, \mathcal{D}_{ij})$ is a $DTS(14,3)$.

Direct calculation shows that each $\mathcal{D}_{ij} (i \in Z_3, j \in I_4)$ contains 182 transitive triples, just the number expected. Therefore we only need to show that every ordered pair P of X appears in three transitive triples of $\mathcal{D}_{ij}$. All the possibilities are exhausted as follows (we only give the proof for $\mathcal{D}_{0j}$, the proof for $\mathcal{D}_{1j}$, or $\mathcal{D}_{2j}$ is similar):

(i) $P = (a,b)$ or (b,a). P appears in three triples of part (1).

(ii) $P = (a,(k,x)), k \in Z_3$ and $x \in I_4$. P appears in one transitive triple of part (1) and in two triples of part (4). Similar for $P = ((k,x),a), (b,(k,x))$ and $((k,x),b)$.

(iii) $P = ((k,x),(k+1,y)), k \in Z_3$ and $x,y \in I_4$. There exists $v \in I_4$ such that $y = (x \circ v)\alpha^j$. If $x \neq v$, P is contained in one triple of part (2) and in two triples of part (3). If $x = v$, P is contained in one triple of part (2) and in two triples of part (4). It is similar for $P = ((k,x),(k-1,y))$.

(iv) $P = ((k,x),(k,y)), k \in Z_3$ and $x \neq y \in I_4$. P is contained in one transitive triple of part (1) and in two triples of part (3).

2°. $\{(X, \mathcal{D}_{ij}) : i \in Z_3, j \in I_4\}$ is an $LDTS(14,3)$.

We only have to proved that every transitive triple T of X is contained in some $\mathcal{D}_{ij}$. All the possibilities are exhibited as follows:

(i) $T = ((i,x),(i,y),(i,z)), i \in Z_3$ and x,y,z are distinct elements of X (whenever a or b appears for x,y,z, omit the first coordinate i). There exists $j \in I_4$ such that $(x,y,z) \in \mathcal{B}_{ij}$, T is contained in part (1) of $\mathcal{D}_{ij}$.

(ii) $T = (a,(k,x),(l,y)), k \neq l \in Z_3$ and $x, y \in I_4$. There exists $j \in I_4$ such that $y = x\alpha^j$. If $l = k+1$, T is contained in (4) of $\mathcal{D}_{0j}$. If $l = k+2$, T is contained in (4) of $\mathcal{D}_{2j}$. (It is similar for $T = ((k,x), a, (l,y))$, $((k,x), (l,y), a)$, $(b,(k,x), (l,y))$, $((k,x), b, (l,y))$ and $((k,x), (l,y), b)$).

(iii) $T = ((i,x),(i+1,y),(i+2,z))$ or $((i+2,z),(i+1,y),(i,x)), i \in Z_3$ and $x, y, z \in I_4$. There exists $j \in I_4$ such that $z = (x \circ y)\alpha^j$. T is contained in (2) of $\mathcal{D}_{ij}$.

(iv) $T = ((i,x),(i,y),(i+1,z)), i \in Z_3$ and $x, y, z \in I_4$ and $x \neq y$. There exists $j \in I_4$ such that $z = (x \circ y)\alpha^j$, thus T is contained in (3) of $\mathcal{D}_{0j}$. It is similar for other five cases. □

Theorem 4.4.5. There exists an $LDTS(v,3)$ for any $v \neq 2$.

Proof. When $v \neq 6$, 14, 62, by Lemma 4.4.2 and Theorem 4.4.3 there exists an $LDTS(v,3)$. By Lemma 4.4.4 there exists an $LDTS(14,3)$. Applying Theorem 4.4.1 with $n = 5, m = 12$, an $LDTS(62,3)$ exists. And an $LDTS(6,3) = \{(I_6, \mathcal{D}_j)\colon\ j \in I_4\}$ can be obtained from an $LDTS(6,1) = \{(I_6, \mathcal{B}_{ij}) : i \in I_3, j \in I_4\}$, where $\mathcal{D}_j = \bigcup_{i \in I_3} \mathcal{B}_{ij}$, $j \in I_4$. □

4.4.2 Completion of spectrum

Theorem 4.4.6. There exists an $LDTS(v,\lambda)$ if and only if either $\lambda \equiv 1,2 \pmod 3$ and $v \equiv \lambda+2, 2\lambda+2 \pmod{3\lambda}$ or $\lambda \equiv 0 \pmod 3$ and $v \equiv 2 \pmod{\frac{\lambda}{3}}$, $v \neq 2$

Proof. The necessary of the existence of $LDTS(v,\lambda)$ is

either $\lambda \equiv 1,2 \pmod 3$, $v \equiv 0,1 \pmod 3$ and $\lambda | 3(v-2)$

or $\lambda \equiv 0 \pmod 3$ and $\lambda | 3(v-2)$ and $v \neq 2$.

These conditions are equivalence the conditions of

either $\lambda \equiv 1,2 \pmod 3$ and $v \equiv \lambda+2, 2\lambda+2 \pmod{3\lambda}$

or $\lambda \equiv 0 \pmod 3$, $v \equiv 2 \pmod{\frac{\lambda}{3}}$ and $v \neq 2$.

Now we only need to show that these conditions are sufficient. When $\lambda \equiv 1,2 \pmod 3$, $v \equiv \lambda+2, 2\lambda+2 \pmod{3\lambda}$, then $\lambda | 3(v-2)$ and $v \equiv 0,1 \pmod 3$. By Theorem 4.3.1, there exists an $LDTS(v,1)$. Let X be a

v–set, $t = \frac{3(v-2)}{\lambda}$, and let

$$LDTS(v,1) = \{(X, \mathcal{B}_i) : i = 1, 2, ..., 3(v-2)\}.$$

Define

$$\mathcal{D}_j = \bigcup_{i=0}^{\lambda-1} \mathcal{B}_{it+j}, \ j = 1, 2, ..., t.$$

Then it is easy to check that $\{(X, \mathcal{D}_j) : j = 1, 2, ..., \frac{3(v-2)}{\lambda}\}$ is an $LDTS(v, \lambda)$.

When $\lambda \equiv 0 \pmod 3$, $v \equiv 2 \pmod{\frac{\lambda}{3}}$ and $v \neq 2$, then $\frac{\lambda}{3} | v - 2$. By Theorem 4.4.5, there exists an $LDTS(v, 3)$. Let X be a v–set, $t = \frac{3(v-2)}{\lambda}$, and let

$$LDTS(v,3) = \{(X, \mathcal{B}_i) : i = 1, 2, ..., v-2\}.$$

Set

$$\mathcal{D}_j = \bigcup_{i=0}^{\frac{\lambda}{3}-1} \mathcal{B}_{it+j}, \ j = 1, 2, ..., t.$$

Then $\{(X, \mathcal{D}_j) : j = 1, 2, ..., \frac{3(v-2)}{\lambda}\}$ is an $LDTS(v, \lambda)$. □

Chapter 5

Large sets of hybrid triple systems

5.1 Introduction

Let X be a v-set. There are two kinds of oriented triple (underlied in a 3-subset $\{x, y, z\}$ of X):

cyclic triple, which contains ordered pairs $(x, y), (y, z)$ and (z, x);

transitive triple, which contains ordered pairs $(x, y), (y, z)$ and (x, z).

An *oriented triple system* is a pair $(X, \mathcal{B})$, where $\mathcal{B}$ is a collection of oriented triples such that each ordered pair of distinct elements of X is contained in exactly λ triples of $\mathcal{B}$. If $\mathcal{B}$ consists of cyclic triples only, the system is called *Mendelsohn triple system* and denoted by $MTS(v, \lambda)$. If $\mathcal{B}$ consists of transitive triples only, the system is called *directed triple system* and denoted by $DTS(v, \lambda)$. But, if there are both cyclic and transitive triples in $\mathcal{B}$, then the system is called *hybrid triple system* and denoted by $HTS(v, \lambda)$. C. J. Colbourn, W. R. Pulleyblank and A. Rosa first introduced this generalization of MTS and DTS in 1989.

For the existence of HTS, the way of saying it is what flexibility is there in choosing the number of cyclic triples in the system ? For it, the concept c-HTS was defined in [ColPulRo]. By a c-$HTS(v, \lambda)$ one means that there are just c cyclic triples in $\mathcal{B}$. Since a $HTS(v, \lambda)$ contains $\lambda v(v-1)/3$ triples, thus $0 \leq c \leq \lambda v(v-1)/3$. But, a c-$HTS(v, \lambda)$

will become a $DTS(v,\lambda)$ when $c=0$, or become an $MTS(v,\lambda)$ when $c=\lambda v(v-1)/3$. In this Chapter, we will only consider the cases $0<c<\lambda v(v-1)/3$, i.e. genuine hybrid triple system. In what follows, an $HTS(v,\lambda)$ means a c-$HTS(v,\lambda)$ for certain $c\in[1,\lambda v(v-1)/3-1]$.

By [ColPulRo] and [Hein], the spectrum of c-$HTS(v,\lambda)$ is $v\geq 3$, $1\leq c\leq \frac{\lambda v(v-1)}{3}-2$, if $\lambda\not\equiv 0 \pmod 3$ then $v-2\not\equiv 0 \pmod 3$ and if $\lambda=1$ and $v=6$ then $c\neq 8$. Note that there exists no $HTS(3,1)$, since an oriented triple system on the 3-set $\{a,b,c\}$ has only two types: $\{\langle a,b,c\rangle,\langle c,b,a\rangle\}$ (i.e. $MTS(3)$) and $\{(a,b,c),(c,b,a)\}$ (i.e. $DTS(3)$).

As examples, for $1\leq c\leq 7$, we give seven c-$HTS(6,1)$s as follows:

1-$HTS(6,1)$	2-$HTS(6,1)$	3-$HTS(6,1)$
$\langle 1,4,3\rangle(0,1,3)$	$\langle 1,2,5\rangle(1,0,4)$	$\langle 1,2,5\rangle(0,1,3)$
$(0,4,5)$	$\langle 2,4,3\rangle(1,3,5)$	$\langle 1,4,3\rangle(1,0,5)$
$(1,2,0)$	$(2,0,3)$	$\langle 2,3,5\rangle(2,0,4)$
$(2,1,5)$	$(3,0,1)$	$(3,0,2)$
$(2,3,4)$	$(4,0,5)$	$(4,2,1)$
$(3,5,2)$	$(4,2,1)$	$(4,5,0)$
$(4,0,2)$	$(5,0,2)$	$(5,3,4)$
$(5,3,0)$	$(5,3,4)$	
$(5,4,1)$		

4-$HTS(6,1)$	5-$HTS(6,1)$	6-$HTS(6,1)$
$\langle 1,2,4\rangle(0,1,5)$	$\langle 0,1,3\rangle(0,4,5)$	$\langle 0,1,2\rangle(1,3,5)$
$\langle 1,3,5\rangle(0,3,2)$	$\langle 1,4,3\rangle(1,2,0)$	$\langle 0,2,3\rangle(2,5,4)$
$\langle 1,4,3\rangle(2,1,0)$	$\langle 2,1,5\rangle(2,3,4)$	$\langle 0,3,4\rangle(4,2,1)$
$\langle 2,5,4\rangle(3,0,4)$	$\langle 2,5,3\rangle(4,0,2)$	$\langle 0,4,5\rangle(5,3,2)$
$(4,5,0)$	$\langle 3,5,0\rangle(5,4,1)$	$\langle 0,5,1\rangle$
$(5,2,3)$		$\langle 1,4,3\rangle$

7-$HTS(6,1)$
$\langle 0,2,5\rangle$ $(0,1,4)$
$\langle 0,3,4\rangle$ $(2,3,0)$
$\langle 0,5,1\rangle$ $(4,5,2)$
$\langle 1,2,4\rangle$
$\langle 1,3,2\rangle$
$\langle 1,5,3\rangle$
$\langle 3,5,4\rangle$

If all cyclic (resp. transitive) triples from X can be partitioned into some $\mathcal{B}_i$ such that each $(X, \mathcal{B}_i)$ is an $MTS(v, \lambda)$ (resp. $DTS(v, \lambda)$), then $\{(X, \mathcal{B}_i)\}_i$ is called a *large set* of pairwise disjoint Mendelsohn (resp. directed) triple systems and denoted by $LMTS(v, \lambda)$ (resp. $LDTS(v, \lambda)$). If all cyclic and transitive triples from X can be partitioned into some $\mathcal{B}_i$ such that each $(X, \mathcal{B}_i)$ is an c_i-$HTS(v, \lambda)$, then $\{(X, \mathcal{B}_i)\}_i$ is called a *large set* of pairwise disjoint hybrid triple systems and denoted by $LHTS(v, \lambda)$.

The existence spectrum of $LMTS(v, \lambda)$ and $LDTS(v, \lambda)$ have been completed (see Chapter 3 and Chapter 4). In this Chapter, we will discuss the existence of $LHTS(v, \lambda)$.

5.2 Case $\lambda = 1$

It is easy to see that the necessary condition for the existence an $LHTS(v, 1)$ is $v \equiv 0,\ 1 \pmod 3$, since there is no $HTS(3, 1)$. And each $LHTS(v, 1)$ consists of $4(v-2)$ $HTS(v, 1)$s. Obviously, there is no $LHTS(3, 1)$.

Lemma 5.2.1. There exist an $LHTS(4, 1)$ and an $LHTS(6, 1)$.

Construction. (1) We can give two non-isomorphic $LHTS(4, 1)$. The first one is $LHTS(4) = \{(Z_4, \mathcal{B}_{ij}) : 0 \le i \le 1, j \in Z_4\}$, where

$$\begin{aligned}
\mathcal{B}_{00} &= \{\langle 1,2,3\rangle, (1,0,3), (2,0,1), (3,0,2)\};\\
\mathcal{B}_{0j} &= \mathcal{B}_{00} + j, j \in Z_4;\\
\mathcal{B}_{1j} &= \mathcal{B}_{0j}^{-1},\ j \in Z_4.
\end{aligned}$$

Here, the system $\mathcal{B}_{0j}^{-1}$ consists of the reversal triples of all triples in $\mathcal{B}_{0j}$, where the reverse of cyclic triple $\langle x, y, x\rangle$ is $\langle z, y, x\rangle$ and the reverse of transitive triple (x, y, z) is (z, y, x). Note that each $\mathcal{B}_{ij}$ contains one cyclic triple and three transitive triples. And all $\mathcal{B}_{0j}$ (resp. $\mathcal{B}_{1j}$) are isomorphic each other.

As well as, replacing $\mathcal{B}_{00}$ by

$$\mathcal{B}'_{00} = \{\langle 1,2,3\rangle, (0,2,1), (1,0,3), (3,2,0)\},$$

we can obtain the second

$$LHTS(4) = \{(Z_4, \mathcal{B}'_{ij}) : 0 \le i \le 1, j \in Z_4\},$$

which is non-isomorphic to the above one.

(2) In Table 2 of [LinSt 2], an $LHTS(6,1)$ was given. Here, we present another construction, which is non-isomorphic to preceding one:

$LHTS(6) = \{(Z_6, \mathcal{B}_{ij}) : 0 \le i \le 1, j \in Z_6\} \bigcup \{(Z_6, \mathcal{C}_{ik}) : 0 \le i, k \le 1\}$, where

$\mathcal{B}_{00} = \{\langle 1,4,3\rangle, (0,1,3), (0,4,5), (4,0,2), (1,2,0), (5,3,0), (5,4,1), (3,5,2), (2,1,5), (2,3,4)\}$;

$\mathcal{B}_{0j} = \mathcal{B}_{00} + j, j \in Z_6$;

$\mathcal{B}_{1j} = \mathcal{B}_{0j}^{-1}, j \in Z_6$;

$\mathcal{C}_{00} = \{\langle 1,5,3\rangle, \langle 0,2,5\rangle, \langle 2,4,1\rangle, \langle 4,0,3\rangle, \langle 0,5,1\rangle, \langle 2,1,3\rangle, \langle 4,3,5\rangle, (0,1,4), (2,3,0), (4,5,2)\}$;

$\mathcal{C}_{01} = \mathcal{C}_{00} + 1$;

$\mathcal{C}_{1j} = \mathcal{C}_{0j}^{-1}, \ j = 0, 1.$

Here, each $\mathcal{B}_{ij}$ contains one cyclic triple and fourteen transitive triples; each $\mathcal{C}_{ij}$ contains seven cyclic triples and three transitive triples.

Lemma 5.2.2. There exists an $LHTS(7,1)$.

Construction. Let set $X = \{\infty_1, \infty_2\} \bigcup Z_5$. For any 3-subset $B = \{x, y, z\}$ of X, define

$$\sigma_0(B) = \{\langle x, y, z\rangle, \langle x, z, y\rangle\}, \ \sigma_1(B) = \{(x, y, z), (z, y, x)\},$$

$$\sigma_2(B) = \{(x, z, y), (y, z, x)\}, \ \sigma_3(B) = \{(y, x, z), (z, x, y)\}.$$

And define

$$\tau_0 = \{\langle \infty_1, 1, 3\rangle, \langle \infty_1, 3, 4\rangle, \langle \infty_1, 4, 2\rangle, \langle \infty_1, 2, 1\rangle\},$$

$$\tau_1 = \{(\infty_1, 1, 3), (3, 4, \infty_1), (\infty_1, 4, 2), (2, 1, \infty_1)\},$$

$$\tau_2 = \{(1, \infty_1, 3), (3, \infty_1, 4), (4, \infty_1, 2), (2, \infty_1, 1)\},$$

$$\tau_3 = \{(1, 3, \infty_1), (\infty_1, 3, 4), (4, 2, \infty_1), (\infty_1, 2, 1)\}.$$

But each τ_i', $0 \le i \le 3$, consists of the reversal blocks of τ_i, where ∞_1 replaces with ∞_2. Let

$$\mathcal{B}_0^0 = \sigma_0(A) \bigcup \tau_1 \bigcup \tau_2' \bigcup \sigma_3(B) \bigcup \sigma_3(C),$$

$$\mathcal{B}_0^1 = \sigma_1(A) \bigcup \tau_2 \bigcup \tau_3' \bigcup \sigma_0(B) \bigcup \sigma_0(C),$$

$$\mathcal{B}_0^2 = \sigma_2(A) \bigcup \tau_3 \bigcup \tau_0' \bigcup \sigma_1(B) \bigcup \sigma_1(C),$$

$$\mathcal{B}_0^3 = \sigma_3(A) \bigcup \tau_0 \bigcup \tau_1' \bigcup \sigma_2(B) \bigcup \sigma_2(C),$$

where $A = \{\infty_1,\ \infty_2,\ 0\}$, $B = \{0,\ 1,\ 4\}$ and $C = \{0,\ 2,\ 3\}$. Let

$$\mathcal{B}_i^k = \{B + i \pmod 5) :\ B \in \mathcal{B}_0^k\},\ i \in Z_5,\ k \in Z_4.$$

Then $\{(\{\infty_1,\ \infty_2\} \bigcup Z_5,\ \mathcal{B}_i^k) :\ i \in Z_5,\ k \in Z_4\}$ is an $LHTS(7,1)$. □

Theorem 5.2.3. There exists an $LHTS(v,1)$ if and only if $v \equiv 0,\ 1 \pmod 3$ and $v > 3$.

Proof. The necessity is obviously. By Lemma 5.2.2, there exists an $LHTS(7,\ 1)$. But for $v > 3$, $v \ne 7$ and $v \equiv 0,\ 1 \pmod 3$, by Lemma 4.3.2 (from Proposition 4.4 in [Teir 8]), there exists an $LS(2,(3,\{3,4,6\}),v) = \{(X,\ \mathcal{B}_r) :\ r \in R\}$, where X is a v-set and $|R| = v - 2$. For each $B \in \bigcup_{r \in R} \mathcal{B}_r$, B appears in $|B| - 2$ systems $\mathcal{B}_r$:

If $|B| = 3$ and $B \in \mathcal{B}_r$, denote $LMTS(3,\ 1) = \{(B,\ C_B(r,\ 0))\}$ and $LDTS(3,1) = \{(B,\ C_B(r,i)) :\ i \in I_4 \setminus \{0\}\}$, where $I_4 = \{0,\ 1,\ 2,\ 3\}$;

If $|B| = 4$ and $B \in \mathcal{B}_{r_j}$ (j=1, 2), denote $LHTS(4,1) = \{(B,\ H_B(r_j,\ i)) :\ j = 1,\ 2,\ i \in I_4\}$ by Lemma 5.2.1;

If $|B| = 6$ and $B \in \mathcal{B}_{r_j}$ (j=1, 2, 3, 4), denote $LHTS(6,1) = \{(B,\ H_B(r_j,\ i)) :\ i \in I_4 \text{ and } j = 1,\ 2,\ 3,\ 4\}$ by Lemma 5.2.1.

Define

$$\mathcal{A}_r^i = \Big(\bigcup_{B \in \mathcal{B}_r,\ |B|=3} C_B(r,\ i)\Big) \bigcup \Big(\bigcup_{B \in \mathcal{B}_r,\ |B|=4,6} H_B(r,\ i)\Big),$$

$r \in R$, $i \in I_4$. It is not difficult to verify that each ordered pair of distinct elements of X is contained in exactly one block of $\mathcal{A}_r^i$.

When v is even, i.e. $v \equiv 0,\ 4 \pmod 6$, each $\mathcal{B}_r$ must contain a block of length 4 or 6. Thus, each $\mathcal{A}_r^i$ is an $HTS(v,1)$, certainly. Thereby, $\{(X,\ \mathcal{A}_r^i) : r \in R,\ i \in I_4\}$ is an $LHTS(v,1)$.

When v is odd, i.e. $v \equiv 1,\ 3 \pmod 6$, the $\mathcal{B}_r$ maybe contains triples only, therefore, the oriented triple system $C_B(r,i)$ is not hybrid. In the case, we have to make some change for the structure of $\mathcal{A}_r^i$: If some $\mathcal{B}_r$ consists of triples only, then it contains at least four triples (note that v≥ 7). Take $B_i \in \mathcal{B}_r$, $|B_i| = 3$ and i=0, 1, 2, 3. In $\mathcal{A}_r^i$, replace $\bigcup_{j=1}^{4} C_{B_j}(r,\ i)$ by $\bigcup_{j=1}^{4} C_{B_j}(r,\ i+j)$. After the modification, it is easy to see that each $\mathcal{A}_r^i$ will also be an $HTS(v,1)$.

This completes the proof. □

5.3 Case $\lambda = 3$

There exists an $LHTS(v,3)$ only if $v \equiv 2 \pmod 3$ and $v \geq 5$, since it contains $\frac{4(v-2)}{3}$ disjoint $HTS(v,3)$s.

Lemma 5.3.1. There exists an $LHTS(5,3)$.

Construction. On the set X=$\{\infty\} \bigcup Z_4$ construct the following four $\mathcal{B}_i$, which will give an $LHTS(5,\ 3)$=$\{(X, \mathcal{B}_i) :\ i \in Z_4\}$:

part 1. $MTS(3,1)$ on the set $Z_4 \setminus \{i\}$;

part 2. $\langle \infty, i, x\rangle$ and (x, i, ∞), where $x \in Z_4 \setminus \{i\}$;

part 3. $(i+1, \infty, i+2)$, $(i+2, \infty, i+1)$, $(i+3, \infty, i+1)$, $(\infty, i+1, i+3)$, $(\infty, i+2, i+3)$, $(\infty, i+3, i+2)$;

part 4. $(i, i+1, i+2)$, $(i+3, i, i+2)$, $(i+1, i+3, i)$, $(i, i+3, i+1)$, $(i+2, i, i+3)$, $(i+2, i+1, i)$.

Lemma 5.3.2. There exists an $LHTS(8,3)$.

Construction. Let X=$\{\infty_0,\ \infty_1\} \bigcup (Z_2 \times Z_3)$. For $x \in Z_3$, on the set $\{\infty_0, \infty_1, (0,\ x), (1, x)\}$ define $LHTS(4,\ 1)$=$\{C_r(x)\colon\ 1 \leq r \leq 8\}$. Denote $C_r = \bigcup\{C_r(x)\colon\ x \in Z_3\}$. For $x \neq y \in Z_3$, on the set $\{(0,x), (1,x), (0,y), (1,\ y)\}$, define $LHTS(4,\ 1)$= $\{D_r(x,y)\colon\ 1 \leq r \leq 8\}$. Denote $D_r = \bigcup\{D_r(x,y)\colon\ x \neq y \in Z_3\}$. For $i, j, k \in Z_2$, define $LMTS(3,$

1)=$\{M(i,j,k)\}$ and $LDTS(3,\ 1)$= $\{N_s(i,j,k)\ :\ 2 \le s \le 4\}$ on the set $\{(i,0),(j,1),(k,2)\}$. For $t \in Z_2$, denote

$$E_{1+4t} = \bigcup\{M(i,j,k)\ :\ i+j+k=t\} \text{ and}$$

$$E_{s+4t} = \bigcup\{N_s(i,j,k)\ :\ i+j+k=t\},\ 2 \le s \le 4.$$

For $x, y \in Z_3$, $x < y$ and $t \in Z_2$, define

$F_{1+4t}(x,y) = \{\langle\infty_t,\ (i,x),\ (i,y)\rangle,\ \langle\infty_{1-t},\ (i,x),\ (1-i,y)\rangle$ and their reversal: $i \in I_2\}$;

$F_{2+4t}(x,y) = \{(\infty_t,\ (i,x),\ (i,y)),\ (\infty_{1-t},\ (i,x),\ (1-i,y))$ and their reversal: $i \in I_2\}$;

$F_{3+4t}(x,y) = \{((i,x),\ \infty_t,\ (i,y)),\ ((i,x),\ \infty_{1-t},\ (1-i,y))$ and their reversal: $i \in I_2\}$;

$F_{4+4t}(x,y) = \{((i,x),\ (i,\ y),\ \infty_t),\ ((i,x),\ (1-i,y),\ \infty_{1-t})$ and their reversal: $i \in I_2\}$.

Denote $F_r = \bigcup\{F_r(x,y)\ :\ x,y \in Z_3 \text{ and } x<y\}$. Let

$$\mathcal{B}_r = C_r \bigcup D_r \bigcup E_r \bigcup F_r,\ 1 \le r \le 8.$$

Then $\{(X,\ \mathcal{B}_r)\ :\ 1 \le r \le 8\}$ is an $LHTS(8,\ 3)$.

Lemma 5.3.3. If there exists an $LHTS(u+2,1)$ and $u \ge 3$, then there exists an $LHTS(3u+2,3)$.

Construction. Let $(Q,\circ)$ be an idempotent quasigroup of order u, Z_3 be the residue class ring modulo 3, $\infty_1,\ \infty_2 \notin Z_3 \times Q$ and α be a cyclic permutation of length u over Q. Construct $4u$ oriented triple systems on the set $X = \{\infty_1,\ \infty_2\} \bigcup (Z_3 \times Q)$:

$$\mathcal{A}_r^k = B(r,k) \bigcup C(r,k) \bigcup D(r,k) \bigcup E(r,k),\ r \in Q,\ k \in I_4.$$

The structures of $B(r,k)$, $C(r,k)$, $D(r,k)$ and $E(r,k)$ are as follows:

For $i \in Z_3$, denote $LHTS(u+2,\ 1)$=$\{B_r^k(i)\ :\ r \in Q,\ k \in I_4\}$ on the set $\{\infty_1,\ \infty_2\} \bigcup (\{i\} \times Q)$. Define $B(r,k) = \bigcup_{i \in Z_3} B_r^k(i)$.

Define

$C(r,0) = \{\langle(i,x),\ (i,y),\ (i \pm 1,\ (x \circ y)\alpha^r)\rangle\ :\ x \ne y \in Q,\ i \in Z_3\}$;

$C(r,1) = \{((i,x),\ (i,y),\ (i \pm 1,\ (x \circ y)\alpha^r))\ :\ x \ne y \in Q,\ i \in Z_3\}$;

$C(r,2) = \{((i,x),\ (i \pm 1,\ (x \circ y)\alpha^r),\ (i,y)) :\ x \neq y \in Q,\ i \in Z_3\};$
$C(r,3) = \{((i \pm 1, (x \circ y)\alpha^r),\ (i,x),\ (i,y)) :\ x \neq y \in Q,\ i \in Z_3\}.$
$D(r,0) = \{\langle \infty_t,\ (i,x),\ (i \pm 1,\ x\alpha^r)\rangle :\ x \in Q,\ i \in Z_3,\ t \in \{1,2\}\};$
$E(r,0) = \{\langle (0,x),\ (1,y),\ (2,(x \circ y)\alpha^r)\rangle,\ \langle (0,x),\ (2,y),$
$(1,(x \circ y)\alpha^r)\rangle :\ x,y \in Q\}.$

For $k \in I_4 \setminus \{0\}$, define

$$D(r,k) = \{(\infty_t,\ (k,x),\ (k \pm 1, x\alpha^r)),\ ((k \pm 1,x),\ \infty_t,\ (k \pm 2, x\alpha^r)),$$

$$((k \pm 2,x),\ (k,x\alpha^r),\ \infty_t) :\ x \in Q,\ t = 1,2\}.$$

$$E(r,k) = \{((k,x),\ (k+1,y),\ (k+2,(x \circ y)\alpha^r)),\ ((k+2,(x \circ y)\alpha^r),$$

$$(k+1,y),\ (k,x)) :\ x,y \in Q\}.$$

Then $\{(X,\ \mathcal{A}_r^k) :\ r \in Q,\ k \in I_4\}$ is an $LHTS(3u+2,\ 3)$. □

Theorem 5.3.4. There exists an $LHTS(3u+2,3)$ for $u \not\equiv 0$ (mod 3).

Proof. By Theorem 5.2.3, there exists an $LHTS(u+2,1)$ for $u \not\equiv 0$ (mod 3) and u>1. Thus, by Lemma 5.3.3, there exists an $LHTS(3u{+}2,3)$ for u >3 and $u \not\equiv 0$ (mod 3). But, by Lemma 5.3.1 and 5.3.2, there exist an $LHTS(3 \cdot 1+2,\ 3)$ and an $LHTS(3 \cdot 2+2,\ 3)$. Thereby, for any $u \not\equiv 0$ (mod 3) there exists an $LHTS(3u+2,3)$. □

Corollary 5.3.5. There exists an $LHTS(3 \cdot 2^n + 2,\ 3)$ for any non-negative integer n.

In subsection 3.5.3, we have introduced the concept of $LS(2,(3,K_0,K_1,K_2),v)$ and listed some results about it, as Lemma 3.5.6, 3.5.7 and 3.5.8. Here, similarly to the proof of Theorem 3.5.9, we can obtain the following result.

Theorem 5.3.6. There exists an $LHTS(v,3)$ if and only if $v \equiv 2$ (mod 3) and $v \geq 5$.

Construction. Denote $v = 2^n w + 2$, where $n \geq 0$ and $w \equiv 3$ (mod 6). By Lemma 3.5.8, there exists an $LS(2,(3,\{3,4\},\{3\},\{3 \cdot 2^n+2\}),2^n w+2)$=$\{(X,\ \mathcal{B}_r^i) : r \in I_k,\ i \in I_3\}$, where X is a v-set, $I_3 = \{0,1,2\}$, $I_k =$

$\{0, 1, \cdots, k-1\}$ and $k = \frac{2^n w}{3}$. For each $B \in \bigcup\{\mathcal{B}_r^i : r \in I_k, i \in I_3\}$, B appears in $|B| - 2$ systems $\mathcal{B}_r^i$:

If $|B| = 3 \cdot 2^n + 2$, by the definition of $LS(2, (3, K_0, K_1, K_2), v)$, each $\mathcal{B}_r^i$ contains only one block of length $3 \cdot 2^n + 2$, which appears in $3 \cdot 2^n$ systems $\mathcal{B}_r^i$. Thus, we can suppose $B \in \mathcal{B}_{s_t}^i$ ($i \in I_3$, t=1, 2,$\cdots$, 2^n). By Corollary 5.3.5, there exists an $LHTS(3 \cdot 2^n + 2,\ 3)$=$\{(B,\ C_B(s_t, j))$: $j \in I_4, t = 1, 2, \cdots, 2^n\}$.

If $|B|$=4 and $B \in \mathcal{B}_{r_1}^{i_1}$ and $\mathcal{B}_{r_2}^{i_2}$, denote $LHTS(4,\ 1)$= $\{(B,\ D_B(r_t, i_t, m)) : m \in I_4, t = 1, 2\}$;

If $|B|$=3 and $B \in \mathcal{B}_s^t$, denote $LMTS(3, 1) = \{(B, D_B(s, t, 0))\}$ and $LDTS(3, 1) = \{(B, D_B(s, t, j)) : j \in I_4 \setminus \{0\}\}$;

Let A is the unique block of length $3 \cdot 2^n + 2$ in $\mathcal{B}_r^i$. Note that the A in $\mathcal{B}_r^0$, $\mathcal{B}_r^1$ and $\mathcal{B}_r^2$ is same. Define

$$\mathcal{J}_r^m = \left(\bigcup_{i=0}^{2}\{D_B(r, i, m) : B \in \mathcal{B}_r^i, |B| = 3, 4\}\right) \bigcup$$

$$\{C_A(r, m) : A \in \mathcal{B}_r^0, |A| = 3 \cdot 2^n + 2\},$$

$r \in I_k$, $m \in I_4$. Then we get an $LHTS(2^n w + 2,\ 3)$= $\{(X,\ \mathcal{J}_r^m) : r \in I_k, m \in I_4\}$.

Proof. We only need to point out that each $\mathcal{B}_r^i$ is a PBD on X with index 1, but each block A, $|A| = 3 \cdot 2^n + 2$, is replaced by an HTS with index 3, each block B, $|B| = 4$, is replaced by an HTS with index 1 and each block B, $|B| = 3$, is replaced by an MTS or DTS with index 1. Note that, in each $\mathcal{J}_r^m$, there is a sub$-HTS(3 \cdot 2^n + 2, 3)$ certainly, thereby each $\mathcal{J}_r^m$ is hybrid. □

5.4 Exitence spectrum

Theorem 5.4.1. The existence spectrum of $LHTS(v, \lambda)$ is $4(v-2) \geq \lambda$, $4(v-2) \equiv 0 \pmod{\lambda}$, if $\lambda \not\equiv 0 \pmod 3$ then $v \not\equiv 2 \pmod 3$ and if λ=1 then $v \neq 3$.

Proof. An $LHTS(v, \lambda)$ contains $\frac{4(v-2)}{\lambda}$ disjoint $HTS(v,\ \lambda)$, and an $HTS(v,\ \lambda)$ contains $\frac{\lambda v(v-1)}{3}$ oriented triples. Thus, $LHTS(v, \lambda)$ exists only if $\lambda v(v-1) \equiv 0 \pmod 3$ and $4(v-2) \equiv 0 \pmod \lambda$.

For the case $\lambda \equiv 0 \pmod 3$, let $\lambda=3t$, $t\geq 1$, X be a v-set and $4(v-2) \equiv 0 \pmod{3t}$, $4(v-2) \geq 3t$. By Theorem 5.3.6, since $v \equiv 2 \pmod 3$, there exists an $LHTS(v, 3)=\{(X, \mathcal{B}_i)\colon 1 \leq i \leq \frac{4(v-2)}{3}\}$. Define

$$\mathcal{A}_k = \bigcup_{i=kt+1}^{(k+1)t} \mathcal{B}_i, \quad 0 \leq k \leq \frac{4(v-2)}{3t} - 1,$$

then $\{(X, \mathcal{A}_k) : 0 \leq k \leq \frac{4(v-2)}{\lambda} - 1\}$ is an $LHTS(v,\lambda)$.

For the case $\lambda \not\equiv 0 \pmod 3$, let X be a v-set, $v \equiv 0, 1 \pmod 3$, $4(v-2) \equiv 0 \pmod \lambda$ and $v > 3$, $4(v-2) \geq \lambda$. By Theorem 5.2.3, there exists an $LHTS(v,1)= \{(X, \mathcal{B}_i) : 1 \leq i \leq 4(v-2)\}$. Define

$$\mathcal{A}_k = \bigcup_{i=k\lambda+1}^{(k+1)\lambda} \mathcal{B}_i, \quad 0 \leq k \leq \frac{4(v-2)}{\lambda} - 1,$$

then $\{(X, \mathcal{A}_k) : 0 \leq k \leq \frac{4(v-2)}{\lambda} - 1\}$ is an $LHTS(v, \lambda)$.

Lastly, let us consider the case $v = 3$. By the necessary conditions, the index λ only may take 1 or 2 or 4. But, by Theorem 5.2.3, there is no $LHTS(3,1)$. An $LHTS(3, 2)=\{(I_3, \mathcal{B}_i) : i \in I_2\}$ can be gotten as follows:

$$\mathcal{B}_0 = \{\langle 0,1,2\rangle,\ (1,0,2),\ (2,1,0),\ (0,2,1)\};$$

$$\mathcal{B}_1 = \{\langle 0,2,1\rangle,\ (2,0,1),\ (0,1,2),\ (1,2,0)\}.$$

And, the $LHTS(3,4)$ is trivial, which contains one $HTS(3,4) = \{I_3, \mathcal{B}_0 \bigcup \mathcal{B}_1\}$ only. □

Chapter 6

Large sets of Mendelsohn designs

6.1 Introduction

Let X be a set of v elements, and $x_1, x_2, \ldots, x_k$ distinct elements of X, where k is a positive integer and $v \geq k \geq 3$. A *directed k-cycle* is a collection of k ordered pairs (x_1, x_2), $(x_2, x_3), \ldots, (x_{k-1}, x_k)$ and (x_k, x_1), which is denoted as $\langle x_1, x_2, \ldots, x_k \rangle$ or $\langle x_2, x_3, \ldots, x_k, x_1 \rangle \ldots$ or $\langle x_k, x_1, \ldots, x_{k-1} \rangle$. A *Mendelsohn design* $MD(v, k, \lambda)$ on X is a pair $(X, \mathcal{B})$, where $\mathcal{B}$ is a collection of directed k-cycles of X such that each ordered pair of distinct elements of X is covered by exactly λ directed k-cycle of $\mathcal{B}$. The numbers $|X| = v$, k and λ are called the *order*, the *circuit size* and the *index* of the Mendelsohn design $MD(v, k, \lambda)$, respectively. By definition, a Mendelsohn triple system of order v, $MTS(v, \lambda)$, is just an $MD(v, 3, \lambda)$. In this sense, the Mendelsohn system $MD(v, k, \lambda)$ is a generalization of the Mendelsohn triple system $MTS(v, \lambda)$. This type of design was first introduced by D. F. Hsu and A. D. Keedwell in 1985 [Hsu-Keed]. What they defined is a more general concept than $MD(v, k, \lambda)$, i.e. $(v, K, \lambda) - MD$ with block size in a set K.

It is a trivial exercise to see that if $(X, \mathcal{B})$ is an $MD(v, k, \lambda)$, then $|\mathcal{B}| = \frac{\lambda v(v-1)}{k}$. Thus, a necessary condition for the existence of a $MD(v, k, \lambda)$ is

$$(*) \qquad v \geq k \geq 3 \text{ and } \lambda v(v-1) \equiv 0 \pmod{k}.$$

It is well known that the spectrum for $MD(v,3,1)$ is the set of all $v \equiv 0$ or 1 (mod 3), except $v = 6$ [Men]. Concerning the existence of $MD(v,k,1)$ with circuit size $k > 3$, partial results are known. We first discuss the case $k = v$, and then the general case.

(1) Tuscan square.

An *Italian square* is an $n \times n$ array in which each of the symbols $1, 2, \ldots, n$ appears exactly once in each row. A *Latin square* is an Italian square in which each of the symbols also appears exactly once in each column. A *Tuscan square*, besides being Italian, has the property:

For any two symbols a, b, there is at most one row in which b is the symbol immediately to the right of a. It is easy to see that the words "at most" can be replaced by "at least" or "exactly". Sometimes a Tuscan square is also called *row complete*. By the way, a square is *Roman* if it is both Tuscan and Latin.

It is well known that the existence of a Tuscan square with size $v \times v$ is equivalent to the existence of an $MD(v+1, v+1, 1)$. In fact, firstly, it is not difficult to see that the first column and the last column of a Tuscan square consist of pairwise distinct numbers. Suppose A is a $v \times v$ Tuscan square on the set $X = \{1, 2, \ldots, v\}$. Then all rows of the extended array $(\mathbf{0}, A)$ just consist of all directed $(v+1)$-cycles of an $MD(v+1, v+1, 1)$ on the set $X \cup \{0\}$, where $\mathbf{0}$ is a column vector (consisting of v numbers 0). Conversely, if there exists an $MD(v+1, v+1, 1)$ on the set $X \cup \{0\}$, then there are v directed $(v+1)$-cycles in the system and each block contains all of the numbers of $X \cup \{0\}$. Denote these blocks $\langle 0, a_{i1}, a_{i2}, \ldots, a_{iv} \rangle, 1 \leq i \leq v$. Then the array $A = (a_{ij})_1^v$ is just a Tuscan square on the set X.

(2) Decomposition of K_v^* into circuits. A complete symmetric directed graph K_v^* is a simple directed graph with v vertices in which for any two distinct vertices x, y there are arcs $x \to y$ and $y \to x$. In Graph Theory, what is called the problem of the decomposition of K_v^* into circuits is: for what values of v is it possible to partition the arcs of K_v^* into k-circuits (i.e. directed circuits with length k). Obviously, the existence of the decompostion is equivalent to the existence of an $MD(v, k, 1)$. In particular, the existence of a Tuscan square with size $v \times v$ is equivalent to the existence of a Hamilton decomposition of K_{v+1}^*.

It is well known that there exists a Tuscan square of order v if and only if $v \neq 1, 3, 5$ (refer to [Til] and [GoTa]). So we can say that there exists an $MD(v, v, 1)$ if and only if $v \geq 3$ and $v \neq 4, 6$. For the decomposition of K_v^* into k-circuits, our knowledge is less complete. So far, about the non-existence of the $MD(v, k, 1)$, we have seen that there is no $MD(4, 4, 1), MD(6, 6, 1)$ or $MD(6, 3, 1)$. J. C. Bermond conjectured that (in our terminology): for the existence of $MD(v, k, 1)$, the necessary condition $(*)$ is also sufficient except in the cases $v = k = 4, v = k = 6$ and $v = 6, k = 3$. Now, it is has been proved that this conjecture is true for:

(i) $v \equiv 0$ or $1 \pmod{k}$ (J. C. Bermond and V. Faber [BerFa] for k even, D. Sotteau [So] for k odd);

(ii) $4 \leq k \leq 16$ and k even (J. C. Bermond [Ber2] for $k = 10, 12, 14$, J. C. Bermond & V. Faber [BerFa] for $k = 4, 6, 8, 16$);

(iii) v large enough (for a given k) (R. M. Wilson [Wil 3]).

By the condition (i), obviously, if k is a prime power then the conjecture is also true, for example $3 \leq k \leq 13$ and k odd. All that remains for a complete solution of the conjecture is the case $k \geq 15$, where k has at least two distinct prime factors.

Let $C(X)$ be the set of all directed k-cycles of a set X containing v elements, then $|C(X)| = \frac{1}{k}v(v-1)\cdots(v-k+1)$. The following problem arises quite naturally: Given a set X of size v, for which there exists a $MD(v, k, \lambda)$, is it always possible to partition $C(X)$ into $s = \frac{1}{\lambda}(v-2)(v-3)\ldots(v-k+1)$ subsets $\mathcal{B}_1, \mathcal{B}_2, \ldots, \mathcal{B}_s$ so that each of $(X, \mathcal{B}_1), (X, \mathcal{B}_2), \ldots, (X, \mathcal{B}_s)$ is an $MD(v, k, \lambda)$? Such a collection of $MD(v, k, \lambda)$ is called a *large set* of pairwise disjoint $MD(v, k, \lambda)$. We denote it $LMD(v, k, \lambda)$. For $LMD(v, 3, \lambda)$, i.e., $LMTS(v, \lambda)$, its existence spectrum has been obtained (see Chapter 3). However, up to 1989, for the existence of $LMD(v, k, \lambda)$ with circuit size $k > 3$ nothing was known, even though much is known for the $MD(v, k, 1)$. It is easy to see that

an $LMD(n,n,\lambda)$ consists of $\frac{(n-2)!}{\lambda}$ $MD(n,n,\lambda)$s and each $MD(n,n,\lambda)$ contains $\lambda(n-1)$ blocks;

an $LMD(n,n-1,\lambda)$ consists of $\frac{(n-2)!}{\lambda}$ $MD(n,n-1,\lambda)$s and each $MD(n,n-1,\lambda)$ contains λn blocks.

In this chapter we will introduce our work (see [Kang 3], [YangKang]). In Section 6.2 we give construction of $LMD(2t+1,2t+1,1)$, $LMD(2t+1,2t,1)$ and $LMD(2t+2,2t+1,1)$. For p prime, we give another construction for $LMD(p,p,1)$, $LMD(p,p-1,1)$ and $LMD(p+1,p,1)$ in Section 6.3. And, in Section 6.4, some LMD with index 2 will be obtained. In Section 6.5, we give some examples.

6.2 Constructions of some $LMD(v,k,1)$

Let X be a set and $Sym(X)$ the symmetric group on X. For a permutation $\xi \in Sym(X)$ and a directed k-cycle $B = \langle b_1, b_2, \ldots, b_k\rangle$, we denote $B\xi = \langle b_1\xi, b_2\xi, \ldots, b_k\xi\rangle$, where $b_i\xi$ represents the image of the element b_i under the permutation ξ.

Let $\mathcal{B}$ be a set of directed k-cycles on a set X, and let $X' \subseteq X$. A subgroup $G \leq Sym(X)$ fixing $X\backslash X'$ pointwise is called a *complete automorphism group* of $\mathcal{B}$ over X' if the following conditions are satisfied:

1° for any $\xi \in G$ and any $B \in \mathcal{B}$, $B\xi \in \mathcal{B}$;

2° for any B and $B' \in \mathcal{B}$, if there exists $\xi \in Sym(X)$ fixing $X\backslash X'$ pointwise such that $B\xi = B'$, then $\xi \in G$.

Of course, 2° is only possible for $\{B\}\backslash X' = \{B'\}\backslash X'$, where $\{B\}$ represents the set of the elements in the block B.

In this Chapter, $\mathcal{B}$ will be the set of all directed k-cycles of certain $MD(|X|,k,\lambda)$. We shall identify the subgroup of $Sym(X)$ fixing $X\backslash X'$ pointwise with $Sym(X')$.

Lemma 6.2.1. (1) If $(X,\mathcal{B})$ is a $MD(v,k,\lambda)$, then $(X,\mathcal{B}\xi)$ too, where $\xi \in Sym(X)$ and $\mathcal{B}\xi = \{B\xi :\ B \in \mathcal{B}\}$.

(2) If a system $\mathcal{B}$ has a complete automorphism group G over $X' \subseteq X$,

then $\{\mathcal{B}\xi : \xi \in Sym_G(X')\}$ is a collection of pairwise disjoint $\mathcal{B}\xi$, where $Sym_G(X')$ is a set of right coset representatives for G in $Sym(X')$.

(3) If $G = (\sigma)$ is the subgroup generated by σ, a v-cyclic permutation on a v-set X, then $Sym(X\backslash\{x\})$ is a set of right coset representatives for G in $Sym(X)$, for any $x \in X$.

Proof. (1) The permutation ξ on X induces a permutation on $(X \times X)\backslash\Delta_x$, where $\Delta_x = \{(x,x)|x \in X\}$, so, by the definition of the $MD(v,k,1)$, the system $(X, \mathcal{B}\xi)$ is also an $MD(v,k,1)$.

(2) Suppose there exist $B, B' \in \mathcal{B}$ and $\xi \neq \eta \in Sym_G(X')$ such that $B\xi = B'\eta$, then $B(\xi\eta^{-1}) = B'$ and $\xi\eta^{-1} \in G$ by the definition of complete automorphism group G over X'. This implies $G\xi = G\eta$, i.e., ξ and η belong to identical right cosets, which is impossible.

(3) For an arbitrary permutation $\varsigma \in G(\varsigma \neq e, e$ is the unity of $G)$ and any $a \in X, a\varsigma \neq a$ is always true, since $\varsigma = \sigma^i, 1 \leq i \leq v-1$. But for any $\xi \neq \eta \in Sym_G(X\backslash\{x\}), x(\xi\eta^{-1}) = x$ holds. so $\xi\eta^{-1} \notin G$, i.e., $G\xi \neq G\eta$. □

Below, in this and the next section, we will only use the cases $X = X'$ and $X = X' \cup \{\infty\}$.

Theorem 6.2.2. There exists an $LMD(2t+1, 2t+1, 1)$ for any positive integer t.

Constrution. Define $2t$ directed $(2t+1)$-cycles on the set $Z_{2t} \bigcup\{\infty\}$ as follows:

$$B_i = \langle \infty, b_{i0}, b_{i1}, \ldots, b_{i,2t-1} \rangle, \qquad i \in Z_{2t},$$

where $b_{i,2k} = i+k$ and $b_{i,2k+1} = i-k-1$ $(0 \leq k \leq t-1)$. Let $\mathcal{B}_0 = \{B_0, B_1, \ldots, B_{2t-1}\}$ and $\mathcal{B}_j = \mathcal{B}_0\xi_j$, where $\xi_j \in Sym(Z_{2t}\backslash\{0\})$. Then $\{(Z_{2t} \cup \{\infty\}, \mathcal{B}_j) : 0 \leq j \leq (2t-1)!-1\}$ is an $LMD(2t+1, 2t+1, 1)$.

Proof. Since $i+k = i-k'-1$ implies $k+k'+1 = 0$, but $1 \leq k+k'+1 \leq 2t-1$, it follows that each B_i is a directed $(2t+1)-$cycle, indeed.

For $x \in X$, the ordered pair (∞, x) (or (x, ∞)) is covered by the block B_x (or B_{x+t}). For $x \neq y \in X$, the ordered pair (x,y) is covered by the block B_{x+s} (if $y - x = 2s$) or B_{x+s-t} (if $y - x = 2s-1$). Thus, $(Z_{2t} \cup \{\infty\}, \mathcal{B}_0)$ is an $MD(2t+1, 2t+1, 1)$. And, by Lemma 6.2.1(1), each

$(Z_{2t} \cup \{\infty\}, \mathcal{B}_j)$ is also an $MD(2t+1, 2t+1, 1)$.

Furthermore, let the $2t$-cyclic permutation $\sigma = (0, 1, \ldots, 2t-1) \in Sym(Z_{2t})$. Then the subgroup $G = (\sigma)$ is the complete automorphism group of $\mathcal{B}_0$ over Z_{2t}. In fact,

1° For any $\xi \in G$ (let $\xi = \sigma^l$, $0 \le l < 2t$) and $B_i \in \mathcal{B}_0$, we have $B_i\xi = B_i\sigma^l = B_{i+l} \in \mathcal{B}_0$ (subscripts modulo $2t$).

2° For any $B_i, B_j \in \mathcal{B}_0$, if there exists $\xi \in Sym(Z_{2t})$ such that $B_i\xi = B_j$ then it is inevitable that $b_{il}\xi = b_{jl} (0 \le l \le 2t-1)$, since $\infty\xi = \infty$. So, obviously, $\xi = \sigma^{j-i} \in G$. Thus, by Lemma 6.2.1(3), $Sym(Z_{2t} \backslash \{0\}) \subseteq Sym_G(Z_{2t})$. But, we have

$$|Sym_G(Z_{2t})| = \frac{|Sym(Z_{2t})|}{|G|} = \frac{(2t)!}{2t} = (2t-1)! = |Sym(Z_{2t} \backslash \{0\})|.$$

So, $Sym(Z_{2t} \backslash \{0\}) = Sym_G(Z_{2t})$. Note that the number $(2t-1)!$ is just the desired number of $MD(2t+1, 2t+1, 1)$ in an $LMD(2t+1, 2t+1, 1)$. Therefore, by Lemma 6.2.1(2), our construction is all right. □

Theorem 6.2.3. There exists an $LMD(2t+1, 2t, 1)$ for any positive integer $t > 1$.

Construction. Define $2t+1$ directed $2t$-cycles on the set $Z_{2t} \bigcup \{\infty\}$ as follows:

$$B_i = \langle \infty, b_{i0}, b_{i1}, \ldots, b_{i,2t-2} \rangle, \quad i \in Z_{2t},$$
$$C = \langle 0, 1, \ldots, 2t-1 \rangle,$$

where $b_{i,2k} = i + k$, $0 \le k \le t-1$ and $b_{i,2k+1} = i - k - 1$, $0 \le k \le t-2$. Let $\mathcal{B}_0 = \{B_0, B_1, \ldots, B_{2t-1}, C\}$ and $\mathcal{B}_j = \mathcal{B}_0\xi_j$, where $\xi_j \in Sym(Z_{2t} \backslash \{0\}$. Then $\{(Z_{2t} \cup \{\infty\}, \mathcal{B}_j) : 0 \le j \le (2t-1)! - 1\}$ is an $LMD(2t+1, 2t, 1)$.

Proof. Similar to Theorem 6.2.2, we can see that each B_i is indeed a directed $2t$-cycle. As for the block C, it is trivial.

For $x \in Z_{2t}$, the ordered pair (∞, x) (or (x, ∞)) appears in the block B_x (or B_{x-t+1}). For $x \ne y \in Z_{2t}$, the ordered pair (x, y) appears in the block B_{x+s} (if $y - x = 2s$) or B_{x+s-t} (if $y - x = 2s - 1 \ne 1$) or C (if $y - x = 1$). So, $(Z_{2t} \cup \{\infty\}, \mathcal{B}_0)$ is an $MD(2t+1, 2t, 1)$. And, by Lemma 6.2.1 (1), each $(Z_{2t} \cup \{\infty\}, \mathcal{B}_j)$ is also an $MD(2t+1, 2t, 1)$.

Take the $2t$-cyclic permutation $\sigma = (0, 1, \ldots, 2t-1) \in Sym(Z_{2t})$. Then the subgroup $G = (\sigma)$ is a complete automorphism group of $\mathcal{B}_0$ over Z_{2t}. In fact,

1° For any $\xi \in G$ (let $\xi = \sigma^l, 0 \le l < 2t$) and B_i (or C) $\in \mathcal{B}_0$, we have $B_i\xi = B_i\sigma^l = B_{i+l} \in \mathcal{B}_0$ (or $C\sigma = C$).

2° For any $U, V \in \mathcal{B}_0$, if there exists $\xi \in Sym(Z_{2t})$ such that $U\xi = V$, then both U and V are C or not, since $\infty\xi = \infty$. If $U = V = C$, let U and V be the i-th and the j-th shift of $\langle 0, 1, \ldots, 2t-1\rangle$. Then $\xi = \sigma^{j-i} \in G$. If $U = \mathcal{B}_i$ and $V = \mathcal{B}_j$, just like in the proof of Theorem 6.2.2, 2°, we have also $\xi = \sigma^{j-i} \in G$.

Now continue just as in the proof of Theorem 6.2.2. □

Theorem 6.2.4. There exists an $LMD(2t+2, 2t+1, 1)$ for any positive integer t.

Construction. Define $2t+2$ directed $(2t+1)$-cycles on the set $Z_{2t+1} \bigcup \{\infty\}$ as follows:

$$\begin{aligned} B_i &= \langle \infty, b_{i0}, b_{i1}, \ldots, b_{i,2t-1}\rangle, \ i \in Z_{2t+1}, \\ C &= \langle 0, t+1, 1, t+2, 2, \ldots, 2t, t, \rangle, \end{aligned}$$

where $b_{i,2k} = i + k, 0 \le k \le t-1$ and $b_{i,2k+1} = i - k - 1$ (if $0 \le k \le \lfloor \frac{t}{2} \rfloor$ -1) or $i - k - 2$ (if $\lfloor \frac{t}{2} \rfloor \le k \le t-1$). Let $\mathcal{B}_0 = \{B_0, B_1, \ldots, B_{2t}, C\}$ and $\mathcal{B}_j = \mathcal{B}_0\xi_j$, where $\xi_j \in Sym(Z_{2t+1} \backslash \{0\})$. Then $\{(Z_{2t+1} \cup \{\infty\}, \mathcal{B}_j) : 0 \le j \le (2t)! - 1\}$ is an $LMD(2, 2t+1, 2t+2)$.

Proof. First, since the following cases are impossible, it follows that each B_i is indeed a directed $(2t+1)$-cycle:

if $i + k = i - k' - 1$ then $k + k' + 1 = 0$, but $1 \le k + k' + 1 \le t + \lfloor \frac{t}{2} \rfloor - 1$;

if $i + k = i - k'' - 2$ then $k + k'' + 2 = 0$, but $\lfloor \frac{t}{2} \rfloor + 2 \le k + k'' + 2 \le 2t$;

if $i - k' - 1 = i - k'' - 2$ then $k'' - k' = -1$, but $1 \le k'' - k' \le t - 1$.

For $x \in Z_{2t+1}$, th ordered pair (∞, x) (or (x, ∞)) appears in the block B_x (or B_{x+t+1}). For $x \ne y \in Z_{2t+1}$, the ordered pair (x, y) appears in the block B_{x+s} (if $y - x = 2s - 1 > t + 1$ or $y - x = 2s < t + 1$) or B_{x+s-t} (if $y - x = 2s - 1 < t + 1$ or $y - x = 2s > t + 1$) or C (if $y - x = t + 1$).

Take the $(2t+1)$-cyclic permutation $\sigma = (0, 1, \ldots, 2t) \in Sym(Z_{2t+1})$. Then the subgroup $G = (\sigma)$ is a complete automorphism group of $\mathcal{B}_0$ over Z_{2t+1}. In fact, noting that $C\sigma$ is the 2nd shift of C, it can be obtained similarly to the proof of Theorem 6.2.3. The rest of the proof is also similar. □

6.3 Another method for prime case

In this section we will give another method to construct $LMD(p,p,1)$, $LMD(p,p-1,1)$ and $LMD(p+1,p,1)$ for prime p. Their existence has been proved in the last section, but this construction may produce systems that are nonisomorphic to the above ones, since the starting system $\mathcal{B}_0$ has a different complete automorphism group.

Below, F_p will always be a finite field containing p elements $0,1,2,\ldots$, $p-1$ (p is an odd prime), and g is a primitive element of F_p^*. Moreover, $\sigma=(1,g,g^2,\ldots,g^{p-2})$ and $\tau=(0,1,2,\ldots,p-1)$ will represent two fixed permutations in $Sym(F_p)$.

Lemma 6.3.1. (1) The permutations σ and τ generate a subgroup $G=(\sigma,\tau)$ of $Sym(F_p)$. Its order is $p(p-1)$. Each of its elements can be written in the form $\sigma^t\tau^s$ or $\tau^s\sigma^t$, $0\le t\le p-2$, $0\le s\le p-1$. And $\sigma^t\tau^s=\tau^{sg^{-t}}\sigma^t, \tau^s\sigma^t=\sigma^t\tau^{sg^t}$.

(2) For $x\neq y$, $Sym(F_p\backslash\{x,y\})$ is a set of representatives for the right cosets of G in $Sym(F_p)$.

Proof. (1) For an arbitrary element x of F_p, we have

$$\tau\sigma: x\longrightarrow x+1\longrightarrow (x+1)g,$$

$$\sigma\tau^g: x\longrightarrow xg\longrightarrow xg+g=(x+1)g,$$

thus $\tau\sigma=\sigma\tau^g$. Futhermore, for $0\le t\le p-2$ and $0\le s\le p-1$,

$$\tau^s\sigma=\tau^{s-1}(\tau\sigma)=\tau^{s-1}\sigma\tau^g=\tau^{s-2}\sigma\tau^{2g}=\ldots=\sigma\tau^{sg},$$

$$\tau^s\sigma^t=(\tau^s\sigma)\sigma^{t-1}=\sigma\tau^{sg}\sigma^{t-1}=\sigma^2\tau^{sg^2}\sigma^{t-2}=\ldots=\sigma^t\tau^{sg^t},$$

$$\sigma^t\tau^s=\sigma^t\tau^{(sg^{-t})g^t}=\tau^{sg^{-t}}\sigma^t.$$

Nothing that the orders of σ and τ in $Sym(F_p)$ are $p-1$ and p, respectively, we can get all conclusions.

(2) Any permutation $\varsigma\in G$ ($\varsigma\neq e, e$ is unity element of G) fixes one element of F_p at most. But for any $\xi\neq\eta\in Sym(F_p\backslash\{x,y\})$, the permutation $\xi\eta^{-1}\in Sym(F_p\backslash\{x,y\})$ fixes the two elements x,y of F_p. So $\xi\eta^{-1}\notin G$, i.e., $G\xi\neq G\eta$. But, we have

$$|Sym(F_p\backslash\{0,1\})|=(p-2)!=\frac{|Sym(F_p)|}{|G|},$$

so we have found representatives for all right cosets of G. $\square$

Below it will be necessary, in some cases, to distinguish the various forms of the same directed cycle. To this end, for a given directed k-cycle $B = \langle b_0, b_1, \ldots, b_{k-1}\rangle$ we call the same directed cycle $\langle b_m, b_{m+1}, \ldots, b_{k-1}, b_0, b_1, \ldots, b_{m-1}\rangle$ the *m-shift* of B, and denote it by $B^{(m)}, 0 \leq m \leq k-1$.

The construction of $LMD(p,p,1)$.

Take $B_i = \langle 0, g^i, 2g^i, \ldots, (p-1)g^i\rangle, 0 \leq i \leq p-2$. Let $\mathcal{B}_0 = \{B_0, B_1, \ldots, B_{p-2}\}$ and $\mathcal{B}_j = \mathcal{B}_0\xi_j$, where $\xi_j \in Sym(F_p\backslash\{0,1\})$. Then $\{(F_p, \mathcal{B}_j) : 0 \leq j \leq (p-2)! - 1\}$ is an $LMD(p,p,1)$.

Proof. Obviously, each $\mathcal{B}_i$ is a directed p-cycle. For any $x \neq y \in F_p$, let $y - x = g^i$ and $xg^{-i} = m$. Then the ordered pair $(x,y) = (mg^i, (m+1)g^i)$ appears in B_i (at the m-th position). So, $(F_p, \mathcal{B}_0)$ is an $MD(p,p,1)$. And, by Lemma 6.2.1 (1), each $(F_p, \mathcal{B}_j)$ also is.

Furthermore, we point out that the subgroup $G = (\sigma, \tau)$ is a complete automorphism group of $\mathcal{B}_0$ over F_p . In fact,

1° For any $B_i^{(m)} \in \mathcal{B}_0$ and $\xi \in G$, we have $B_i^{(m)}\xi \in \mathcal{B}_0$ since

$$B_0^{(m)}\tau^s = B_0^{(m+s)}, B_0^{(m)}\sigma^t = B_t^{(m)},$$
$$B_i^{(m)}\tau^s = B_0^{(m)}\sigma^i\tau^s = B_0^{(m)}\sigma^{sg^{-i}}\sigma^i = B_0^{(m+sg^{-i})}\sigma^i = B_i^{(m+sg^{-i})},$$
$$B_i^{(m)}\sigma^t = B_0^{(m)}\sigma^{i+t} = B_{i+t}^{(m)}.$$

2° For any $B_i^{(m)}, B_j^{(n)} \in \mathcal{B}_0$, if there exists $\xi \in Sym(F_p)$ such that $B_i^{(m)}\xi = B_j^{(n)}$, then $\xi = \tau^{(n-m)g^i}\sigma^{j-i} \in G$ (by 1°, $B_i^{(m)}\tau^{(n-m)g^i}\sigma^{j-i} = B_i^{(m+(n-m)g^ig^{-i})}\sigma^{j-i} = B_i^{(n)}\sigma^{j-i} = B_{i+(j-i)}^{(n)} = B_j^{(n)}$). Thus, by Lemma 6.3.1(2), $Sym(F_p\backslash\{0,1\})$ is a set of representatives for the right cosets of G in $Sym(F_p)$. Note that the number $(p-2)!$ is just the desired number of $MD(p,p,1)$'s in an $LMD(p,p,1)$, so our construction is all right by Lemma 6.2.1 (2). $\square$

The construction of $LMD(p,p-1,1)$.

Take $B_i = \langle 1+i, g+i, g^2+i, \ldots, g^{p-2}+i\rangle, 0 \leq i \leq p-1$. Let $\mathcal{B}_0 = \{B_0, B_1, \ldots, B_{p-1}\}$ and $\mathcal{B}_j = \mathcal{B}_0\xi_j$, where $\xi_j \in Sym(F_p\backslash\{0,1\})$. Then $\{(F_p, \mathcal{B}_j) : 0 \leq j \leq (p-2)! - 1\}$ is an $LMD(p,p-1,1)$.

Proof. For any $x \neq y \in F_p$, let $(y-x)(g-1)^{-1} = g^m$ and $x - g^m = i$. Then the ordered pair $(x,y) = (g^m + i, g^{m+1} + i)$ appears in B_i (at the

mth position). So, $(F_p, \mathcal{B}_0)$ is an $MD(p, p-1, 1)$. And, by Lemma 6.2.1 (1), each $(F_p, \mathcal{B}_j)$ also is.

To verify that the subgroup $G = \langle \sigma, \tau \rangle$ is a complete automorphism group of $\mathcal{B}_0$ over F_p, we only need to note that

1° For any $B_i^{(m)} \in \mathcal{B}_0$ and $\xi \in G$, we have $B_i^{(m)}\xi \in \mathcal{B}_0$ since
$B_0^{(m)}\tau^s = B_s^{(m)}$, $B_0^{(m)}\sigma^t = B_0^{(m+t)}$ and
$B_i^{(m)}\tau^s = B_0^{(m)}\tau^{i+s} = B_{i+s}^{(m)}$,
$B_i^{(m)}\sigma^t = B_0^{(m)}\tau^i\sigma^t = B_0^{(m)}\sigma^t\tau^{ig^t} = B_0^{(m+t)}\tau^{ig^t} = B_{ig^t}^{(m+t)}$.

2° For any $B_i^{(m)}, B_j^{(n)} \in \mathcal{B}_0$, if there exists $\xi \in Sym(F_p)$ such that $B_i^{(m)}\xi = B_j^{(n)}$, then $\xi = \sigma^{n-m}\tau^{j-ig^{n-m}} \in G$ (by 1°, $B_i^{(m)}\sigma^{n-m}\tau^{j-ig^{n-m}} = B_{ig^{n-m}}^{(n)}\tau^{j-ig^{n-m}} = B_j^{(n)}$).

Now continue just as in the proof of the above construction. □

The construction of $LMD(p+1, p, 1)$.

Take the directed cycles

$$B_{ik} = \langle 1+i, g+i, \ldots, g^k+i, \infty, g^{k+1}+i, \ldots, g^{p-2}+i \rangle,$$

where $0 \le i \le p-1$, $0 \le k \le p-2$, which are obtained from the blocks B_i of the above-mentioned $LMD(p, p-1, 1)$ by inserting a new element ∞ between g^k+i and $g^{k+1}+i$. Moreover, take

$$C_k = \langle 0, g^k(g-1), 2g^k(g-1), \ldots, (p-1)g^k(g-1) \rangle, 0 \le k \le p-2.$$

Let $\mathcal{B}_{0k} = \{B_{0k}, B_{1k}, \ldots, B_{p-1,k}, C_k\}$ and $\mathcal{B}_{jk} = \mathcal{B}_{0k}\xi_j$, where $\xi_j \in Sym(F_p \setminus \{0,1\})$. Then

$$\{(F_p \cup \{\infty\}, \mathcal{B}_{jk}) : \ 0 \le k \le p-2, \ 0 \le j \le (p-2)! - 1\}$$

is an $LMD(p+1, p, 1)$.

Proof. First, for given k, we prove that the system $\mathcal{B}_{0k}$ is an $MD(p+1, p, 1)$.

For any $x \in F_p$, obviously, the ordered pairs (x, ∞) and (∞, x) appear in $B_{x-g^k,k}$ and $B_{x-g^{k+1},k}$ respectively. And for any $x \ne y \in F_p$, let $(y-x)(g-1)^{-1} = g^m$. Then we have

if $m \ne k$, let $x - g^m = i$, then the ordered pair $(x, y) = (g^m + i, g^{m+1} + i)$ appears in B_{ik};

if $m = k$, let $x(g^k(g-1))^{-1} = s$, then the ordered pair $(x,y) = (sg^k(g-1),(s+1)g^k(g-1))$ appears in C_k.
Furthermore, the number of the blocks in $\mathcal{B}_{0k}$ is $p+1$, just as desired.

Let us show that the subgruop $G = (\sigma,\tau)$ is a complete automorphism group over F_p of the collection $\mathcal{B} = \{\mathcal{B}_{0k} : 0 \leq k \leq p-2\}$.

1° For any $U \in \mathcal{B}$,

if $U = C_k^{(m)}$, we have $C_k^{(m)}\tau^s = C_k^{(m+s(g^k(g-1))^{-1})}$ and $C_k^{(m)}\sigma^t = C_{k+t}^{(m)}$;

if $U = B_{ik}$, we have $B_{ik}\tau^s = B_{i+s,k}$ and $B_{ik}\sigma^t = B_{g^t i,k+t}$.

Thus, $U\xi \in \mathcal{B}$ for any $\xi \in G = (\sigma,\tau)$.

2° For any $U,V \in \mathcal{B}$, if there exists $\xi \in Sym(F_p)$ such that $U\xi = V$, then since $\infty\xi = \infty$, there are only two possibilities:

$U = C_k^{(m)}, V = C_l^{(n)}$. Then, by 1°, $C_k^{(m)}\tau^{(n-m)g^k(g-1)}\sigma^{l-k} = C_k^{(n)}\sigma^{l-k} = C_l^{(n)}$, thus $\xi = \tau^{(n-m)g^k(g-1)}\sigma^{l-k}$;

$U = B_{ik}, V = B_{jl}$. Then, by 1°, $B_{ik}\sigma^{l-k}\tau^{j-g^{l-k}i} = B_{g^{l-k}i,l}\tau^{j-g^{l-k}i} = B_{jl})$, thus $\xi = \sigma^{l-k}\tau^{j-g^{l-k}i}$.

In both cases $\xi \in G$.

Now continue just as in the proof of the first construction (Note: in using Lemma 6.2.1 (2), the collection $\mathcal{B}$ is the union of $p-1$ $MD(p+1,p,1)$'s). □

6.4 Construction of some $LMD(v,k,2)$

Let $Z_n = \{0,1,...,n-1\}$ be the residue class ring modulo n and $Sym(Z_n)$ be the symmetric group on Z_n. Let $\alpha = (0,1,...,n-1)$ be a n-cycle on Z_n. And, let

$$\beta = \begin{cases} (0)(\frac{n}{2})(1,n-1)(2,n-2)...(\frac{n}{2}-1,\frac{n}{2}+1) & (n \text{ even}) \\ (0)(1,n-1)(2,n-2)...(\frac{n-1}{2},\frac{n+1}{2}) & (n \text{ odd}) \end{cases}$$

be a permutation on Z_n, too. Define $G = (\alpha,\beta)$ be the subgroup of $Sym(Z_n)$ generalized by α and β. It is well known that G is just the dihedral group, which consists of $2n$ permutations on Z_n. Define $H = (\beta)$ be the subgroup of $Sym(Z_n)$ generalized by β. It is easy to see that

$|H| = 2$ and

$$|Sym_G(Z_n)| = \frac{n!}{2n} = \frac{(n-1)!}{2}, \; |Sym_H(Z_n^*)| = \frac{(n-1)!}{2},$$

where $Z_n^* = Z_n \backslash \{0\}$ and the notation $Sym_G(X)$ is from Lemma 6.2.1(2). It is not hard to verify that each right coset for H in $Sym(Z_n^*)$ is contained one-to-one in certain right coset for G in $Sym(Z_n)$. Thus, $Sym_H(Z_n^*)$ is a set of right coset representatives for G in $Sym(Z_n)$.

Theorem 6.4.1. There exists an $LMD(n+1, n+1, 2)$ for any integer $n \geq 3$.

Construction. Take $X = \{\infty\} \bigcup Z_n$. Define $2n$ directed cycles on X as follows:

$$A_{ij} = \langle \infty, a_{i,1}^j, a_{i,2}^j, \ldots, a_{i,n}^j \rangle, \; i \in Z_n \text{ and } j = 0, 1,$$

where $a_{i,k}^j = i + (-1)^{k+j} \lfloor \frac{k}{2} \rfloor, 1 \leq k \leq n$. Let $\mathcal{B}_0 = \{A_{ij} : i \in Z_n, j = 0, 1\}$ and $\mathcal{B}_l = \mathcal{B}_0 \xi_l$ for $\xi_l \in Sym_H(Z_n^*)$ (or $\xi_l \in Sym_G(Z_n)$). Here, G and H are define above. Then,

$$\{(X, \mathcal{B}_l) : \; 1 \leq l \leq \frac{(n-1)!}{2}\}$$

is an $LMD(n+1, n+1, 2)$ as expected.

Proof. By Lemma 6.2.1(1) and (2), we only need to show the following two facts.

(1) $\mathcal{B}_0$ is an $MD(n+1, n+1, 2)$.

Obviousy, each ordered pair (∞, i) appears twice in $\mathcal{B}_0$, i.e. in blocks A_{i0} and A_{i1}. And, as $\{i + (-1)^n \lfloor \frac{n}{2} \rfloor : i \in Z_n\} = \{i + (-1)^{n+1} \lfloor \frac{n}{2} \rfloor : i \in Z_n\} = Z_n$, each pair (i, ∞) appears twice in $\mathcal{B}_0$, also. Next, let us consider the differences

$$a_{i,k+1}^j - a_{i,k}^j = (-1)^{k+j+1} (\lfloor \frac{k+1}{2} \rfloor + \lfloor \frac{k}{2} \rfloor) = (-1)^{k+j+1} k, \; 1 \leq k \leq n-1.$$

When n is even, $\{(-1)^{k+1} k : \; k \in Z_n^*\} = \{(-1)^{k+2} k : \; k \in Z_n^*\} = Z_n^*$. When n is odd, $\{(-1)^{k+1} k : \; k \in Z_n^*\}$ (resp. $\{(-1)^{k+2} k : \; k \in Z_n^*\}$) is two copies of all odd (resp. even) numbers in Z_n^*. Therefore each ordered pair (x, y), $x \neq y \in Z_n$, appears twice in $\mathcal{B}_0$.

(2) G is a complete automorphism group of $\mathcal{B}_0$ over Z_n.

The subgroup G consists of the following permutations:

$$\alpha^0, \alpha^1, ..., \alpha^{n-1}, \beta, \beta\alpha, \beta\alpha^2, ..., \beta\alpha^{n-1}.$$

We have these facts:

(i) $\alpha^t(a_{i,k}^j) = i + t + (-1)^{k+j}\lfloor\frac{k}{2}\rfloor = a_{i+t,k}^j$,

$\beta\alpha^t(a_{i,k}^j) = -(i + t + (-1)^{k+j}\lfloor\frac{k}{2}\rfloor) = a_{n-i-t,k}^{1-j}$.

Thus, $\alpha^t(A_{ij}) = A_{i+t,j} \in \mathcal{B}_0$ and $\beta\alpha^t(A_{ij}) = A_{n-i-t,1-j} \in \mathcal{B}_0$.

(ii) For any $A_{ij}, A_{st} \in \mathcal{B}_0$, suppose that there be $\xi \in Sym(Z_n)$ such that $A_{ij}\xi = A_{st}$. As $\xi(\infty) = \infty$, thus $\xi(a_{i,k}^j) = a_{s,k}^t$ for $1 \le k \le n$. If $j = t$, then the equations $\xi(i+(-1)^{k+j}\lfloor\frac{k}{2}\rfloor) = s+(-1)^{k+j}\lfloor\frac{k}{2}\rfloor$, $1 \le k \le n$, imply the permutation

$$\xi = \begin{pmatrix} i & i+1 & i-1 & i+2 & i-2 & \cdots \\ s & s+1 & s-1 & s+2 & s-2 & \cdots \end{pmatrix} = \alpha^{s-i} \in G.$$

If $j \ne t$, i.e. $1 - j = t$, then the equations $\xi(i + (-1)^{k+j}\lfloor\frac{k}{2}\rfloor) = s - (-1)^{k+j}\lfloor\frac{k}{2}\rfloor$, $1 \le k \le n$, imply the permutation

$$\xi = \begin{pmatrix} i & i+1 & i-1 & i+2 & i-2 & \cdots \\ s & s-1 & s+1 & s-2 & s+2 & \cdots \end{pmatrix} = \beta\alpha^{-(s+i)} \in G.$$

Therefore, in both cases the conclusion holds.

This completes the proof. □

Theorem 6.4.2. There exists an $LMD(n+1,n,2)$ for any integer $n \ge 3$.

Construction. Take $X = \{\infty\} \bigcup Z_n$. Define $2(n+1)$ directed cycles on X as follows:

$A_{ij} = \langle\infty, a_{i,1}^j, a_{i,2}^j, ..., a_{i,n-1}^j\rangle$, $i \in Z_n$ and $j = 0, 1$,

$C_0 = \langle 0, 1, ..., n-1\rangle$ and $C_1 = \langle n-1, n-2, ..., 1, 0\rangle$,

where $a_{i,k}^j$ is same to last theorem. Let

$$\mathcal{B}_0 = \{C_0, C_1\} \bigcup \{A_{ij} : \ i \in Z_n, j = 0, 1\},$$

and $\mathcal{B}_l = \mathcal{B}_0\xi_l$ for $\xi_l \in Sym_H(Z_n^*)$ (or $Sym_G(Z_n)$). Then

$$\{(X, \mathcal{B}_l) : \ 1 \le l \le \frac{(n-1)!}{2}\}$$

is an $LMD(n+1,n,2)$ as expected.

Proof. Similar to last theorem, we only point out again:

(1) When $k = n - 1$, the difference $(-1)^{k+j+1}k$ is $\pm(n-1) = \pm 1$, which appears in the ordered pairs of C_0 or C_1.

(2) (i) $\alpha^t(C_j) = C_j$ and $\beta(C_j) = C_{1-j}$, for $j = 0, 1$; (ii) Since ∞ is a fixed element under any $\xi \in Sym(Z_n)$, for blocks $U, V \in \mathcal{B}_0$, if $U\xi = V$ then both U and V belong to $\mathcal{B}_0 \backslash \{C_0, C_1\}$ or not. If $U = V = C_j$, then $\xi \in (\alpha) \subset (\alpha, \beta) = G$. If $C_0\xi = C_1$, then $\xi = \beta$ (or $\beta\alpha^t$). □

6.5 Examples

Example 1. $LMD(7,7,1)$.
(1) (by Theorem 6.2.2) $X = \{0,1,2,3,4,5\}$

$$\begin{array}{lll} \mathcal{B}_0: & B_0 & = \langle\infty,0,5,1,4,2,3\rangle \\ & B_1 & = \langle\infty,1,0,2,5,3,4\rangle \\ & B_2 & = \langle\infty,2,1,3,0,4,5\rangle \\ & B_3 & = \langle\infty,3,2,4,1,5,0\rangle \\ & B_4 & = \langle\infty,4,3,5,2,0,1\rangle \\ & B_5 & = \langle\infty,5,4,0,3,1,2\rangle \end{array}$$

$\mathcal{B}_j = \mathcal{B}_0\xi_j, \xi_j \in Sym(X\backslash\{0\}) = S_5$.
(2) (by Section 6.3 $LMD(p,p,1)$)
$p = 7, g = 3, \sigma = (1,3,2,6,4,5), \tau = (0,1,2,3,4,5,6)$.

$$\begin{array}{lll} \mathcal{B}_0: & B_0 & = \langle 0,1,2,3,4,5,6\rangle \\ & B_1 & = \langle 0,3,6,2,5,1,4\rangle \\ & B_2 & = \langle 0,2,4,6,1,3,5\rangle \\ & B_3 & = \langle 0,6,5,4,3,2,1\rangle \\ & B_4 & = \langle 0,4,1,5,2,6,3\rangle \\ & B_5 & = \langle 0,5,3,1,6,4,2\rangle \end{array}$$

$\mathcal{B}_j = \mathcal{B}_0\xi_j, \xi_j \in Sym(F_7\backslash\{0,1\}) = S_5$.

Example 2. $LMD(7,6,1)$. (1) (by Theorem 6.2.3) $X = \{0,1,2,3,4,5\}$

$$\begin{array}{lll} \mathcal{B}_0: & B_0 & = \langle\infty,0,5,1,4,2\rangle \\ & B_1 & = \langle\infty,1,0,2,5,3\rangle \\ & B_2 & = \langle\infty,2,1,3,0,4\rangle \\ & B_3 & = \langle\infty,3,2,4,1,5\rangle \\ & B_4 & = \langle\infty,4,3,5,2,0\rangle \\ & B_5 & = \langle\infty,5,4,0,3,1\rangle \\ & C & = \langle 0,1,2,3,4,5\rangle \end{array}$$

$\mathcal{B}_j = \mathcal{B}_0 \xi_j,\ \xi_j \in Sym(X\backslash\{0\}) = S_5$.

(2) (by Section 6.3 $LMD(p, p-1, 1)$)
p, g, σ, τ are the same as in Example 1 (2).

$$\begin{array}{lll} \mathcal{B}_0: & B_0 & = \langle 1,3,2,6,4,5\rangle \\ & B_1 & = \langle 2,4,3,0,5,6\rangle \\ & B_2 & = \langle 3,5,4,1,6,0\rangle \\ & B_3 & = \langle 4,6,5,2,0,1\rangle \\ & B_4 & = \langle 5,0,6,3,1,2\rangle \\ & B_5 & = \langle 6,1,0,4,2,3\rangle \\ & B_6 & = \langle 0,2,1,5,3,4\rangle \end{array}$$

$\mathcal{B}_j = \mathcal{B}_0 \xi_j,\ \xi_j \in Sym(F_7\backslash\{0,1\}) = S_5$.

Example 3. $LMD(6,5,1)$.
(1) (by Theorem 6.2.4) $X = \{0,1,2,3,4\}$

$$\begin{array}{lll} \mathcal{B}_0: & B_0 & = \langle \infty,0,4,1,2\rangle \\ & B_1 & = \langle \infty,1,0,2,3\rangle \\ & B_2 & = \langle \infty,2,1,3,4\rangle \\ & B_3 & = \langle \infty,3,2,4,0\rangle \\ & B_4 & = \langle \infty,4,3,0,1\rangle \\ & C & = \langle 0,3,1,4,2\rangle \end{array}$$

$\mathcal{B}_j = \mathcal{B}_0 \xi_j,\ \xi_j \in Sym(X\backslash\{0\}) = S_4$.

(2) (by Section 6.3 $LMD(p+1, p, 1)$)
$p = 5, g = 2, \sigma = (1,2,4,3), \tau = (0,1,2,3,4)$

$$\begin{array}{lll} \mathcal{B}_{00}: & B_{00} & = \langle 1,\infty,2,4,3\rangle \\ & B_{10} & = \langle 2,\infty,3,0,4\rangle \\ & B_{20} & = \langle 3,\infty,4,1,0\rangle \\ & B_{30} & = \langle 4,\infty,0,2,1\rangle \\ & B_{40} & = \langle 0,\infty,1,3,2\rangle \\ & C_0 & = \langle 0,1,2,3,4\rangle \end{array} \qquad \begin{array}{lll} \mathcal{B}_{01}: & B_{01} & = \langle 1,2,\infty,4,3\rangle \\ & B_{11} & = \langle 2,3,\infty,4,3\rangle \\ & B_{21} & = \langle 3,4,\infty,1,0\rangle \\ & B_{31} & = \langle 4,0,\infty,2,1\rangle \\ & B_{41} & = \langle 0,1,\infty,3,2\rangle \\ & C_1 & = \langle 0,2,4,1,3\rangle \end{array}$$

$$\begin{array}{lll} \mathcal{B}_{02}: & B_{02} & = \langle 1,2,4,\infty,3\rangle \\ & B_{12} & = \langle 2,3,0,\infty,4\rangle \\ & B_{22} & = \langle 3,4,1,\infty,0\rangle \\ & B_{32} & = \langle 4,0,2,\infty,1\rangle \\ & B_{42} & = \langle 0,1,3,\infty,2\rangle \\ & C_2 & = \langle 0,4,3,2,1\rangle \end{array} \qquad \begin{array}{lll} \mathcal{B}_{03}: & B_{03} & = \langle 1,2,4,3,\infty\rangle \\ & B_{13} & = \langle 2,3,0,4,\infty\rangle \\ & B_{23} & = \langle 3,4,1,0,\infty\rangle \\ & B_{33} & = \langle 4,0,2,1,\infty\rangle \\ & B_{43} & = \langle 0,1,3,2,\infty\rangle \\ & C_3 & = \langle 0,3,1,4,2\rangle \end{array}$$

$\mathcal{B}_{jk} = \mathcal{B}_{0k} \xi_j,\ \xi_j \in Sym(F_5\backslash\{0,1\}) = S_3$.

Example 4. $LMD(6,6,2)$, by Theorem 6.4.1.

$$\begin{array}{llll}
\mathcal{B}_0: & A_{00} = \langle\infty,0,1,4,2,3\rangle, & A_{01} = \langle\infty,0,4,1,3,2\rangle, \\
& A_{10} = \langle\infty,1,2,0,3,4\rangle, & A_{11} = \langle\infty,1,0,2,4,3\rangle, \\
& A_{20} = \langle\infty,2,3,1,4,0\rangle, & A_{21} = \langle\infty,2,1,3,0,4\rangle, \\
& A_{30} = \langle\infty,3,4,2,0,1\rangle, & A_{31} = \langle\infty,3,2,4,1,0\rangle, \\
& A_{40} = \langle\infty,4,0,3,1,2\rangle, & A_{41} = \langle\infty,4,3,0,2,1\rangle.
\end{array}$$

$\mathcal{B}_l = \mathcal{B}_0\xi_l$, $\xi_l \in Sym_H(Z_5^*)$.

Example 5. $LMD(6,5,2)$, by Theorem 6.4.2.

$$\begin{array}{llll}
\mathcal{B}_0: & A_{00} = \langle\infty,0,1,4,2\rangle, & A_{01} = \langle\infty,0,4,1,3\rangle, \\
& A_{10} = \langle\infty,1,2,0,3\rangle, & A_{11} = \langle\infty,1,0,2,4\rangle, \\
& A_{20} = \langle\infty,2,3,1,4\rangle, & A_{21} = \langle\infty,2,1,3,0\rangle, \\
& A_{30} = \langle\infty,3,4,2,0\rangle, & A_{31} = \langle\infty,3,2,4,1\rangle, \\
& A_{40} = \langle\infty,4,0,3,1\rangle, & A_{41} = \langle\infty,4,3,0,2\rangle, \\
& C_0 = \langle 0,1,2,3,4\rangle, & C_1 = \langle 4,3,2,1,0\rangle.
\end{array}$$

$\mathcal{B}_l = \mathcal{B}_0\xi_l$, $\xi_l \in Sym_H(Z_5^*)$.

Here, in Example 4 and 5, the subgroup $H = \{(1),(14)(23)\}$ and the $Sym_H(Z_5^*)$ can be taken as {(1), (12), (13), (14), (24), (34), (123), (124), (134), (12)(34), (132), (1243)}.

Chapter 7

Related designs

7.1 On the existence of golf design

7.1.1 Introduction

A Latin square on the set $X = \{x_1, ..., x_v\}$ is called *symmetric* if the entries in the positions (x_i, x_j) and (x_j, x_i) are same for any $i, j \in \{1, 2, ..., v\}$. Two symmetric Latin squares of order v are called *disjoint* if their entries in the same position in an off–diagonal are different. The set of $v-2$ pairwise symmetric idempotent Latin squares of order v is called a *golf design* of order v, and denoted by $G(v)$. It is easy to know that there exists a $G(v)$ only if v is odd and $v > 1$. First, let us see two examples (see [Wal]).

Example 1. A golf design of order 7, $G(7) = \{(I_7, A_k) : 1 \le k \le 5\}$:

$$A_1 = \begin{bmatrix} 0 & 2 & 4 & 5 & 1 & 6 & 3 \\ 2 & 1 & 6 & 4 & 5 & 3 & 0 \\ 4 & 6 & 2 & 0 & 3 & 1 & 5 \\ 5 & 4 & 0 & 3 & 6 & 2 & 1 \\ 1 & 5 & 3 & 6 & 4 & 0 & 2 \\ 6 & 3 & 1 & 2 & 0 & 5 & 4 \\ 3 & 0 & 5 & 1 & 2 & 4 & 6 \end{bmatrix}, \quad A_2 = \begin{bmatrix} 0 & 3 & 6 & 1 & 5 & 4 & 2 \\ 3 & 1 & 5 & 0 & 6 & 2 & 4 \\ 6 & 5 & 2 & 4 & 1 & 0 & 3 \\ 1 & 0 & 4 & 3 & 2 & 6 & 5 \\ 5 & 6 & 1 & 2 & 4 & 3 & 0 \\ 4 & 2 & 0 & 6 & 3 & 5 & 1 \\ 2 & 4 & 3 & 5 & 0 & 1 & 6 \end{bmatrix},$$

$$A_3 = \begin{bmatrix} 0 & 4 & 5 & 6 & 2 & 3 & 1 \\ 4 & 1 & 3 & 2 & 0 & 6 & 5 \\ 5 & 3 & 2 & 1 & 6 & 4 & 0 \\ 6 & 2 & 1 & 3 & 5 & 0 & 4 \\ 2 & 0 & 6 & 5 & 4 & 1 & 3 \\ 3 & 6 & 4 & 0 & 1 & 5 & 2 \\ 1 & 5 & 0 & 4 & 3 & 2 & 6 \end{bmatrix}, \quad A_4 = \begin{bmatrix} 0 & 5 & 3 & 2 & 6 & 1 & 4 \\ 5 & 1 & 4 & 6 & 3 & 0 & 2 \\ 3 & 4 & 2 & 5 & 0 & 6 & 1 \\ 2 & 6 & 5 & 3 & 1 & 4 & 0 \\ 6 & 3 & 0 & 1 & 4 & 2 & 5 \\ 1 & 0 & 6 & 4 & 2 & 5 & 3 \\ 4 & 2 & 1 & 0 & 5 & 3 & 6 \end{bmatrix},$$

$$A_5 = \begin{bmatrix} 0 & 6 & 1 & 4 & 3 & 2 & 5 \\ 6 & 1 & 0 & 5 & 2 & 4 & 3 \\ 1 & 0 & 2 & 6 & 5 & 3 & 4 \\ 4 & 5 & 6 & 3 & 0 & 1 & 2 \\ 3 & 2 & 5 & 0 & 4 & 6 & 1 \\ 2 & 4 & 3 & 1 & 6 & 5 & 0 \\ 5 & 3 & 4 & 2 & 1 & 0 & 6 \end{bmatrix}.$$

Example 2. A golf design of order 17, $G(17) = \{(Z_{17}, A_k) : 1 \le k \le 15\}$, can be obtained as follows. Let $A_k = (a_{ij}^k)_{i,j \in Z_7}$ and

$$a_{ij}^k = i + b_{k,j-i},\ b_{k,0} = 0,\ b_{k,-j} = b_{k,j} - j,$$

where $1 \le k \le 15, i, j \in Z_{17}$ and all arithmetic being carried out modulo 17. And all parameters $b_{k,j} (1 \le k \le 15, 1 \le j \le 8)$ are taken from the following array $B = (b_{k,j})_{1 \le k \le 15, 1 \le j \le 8}$:

$$B = \begin{bmatrix} 2 & 5 & 9 & 14 & 16 & 13 & 15 & 12 \\ 3 & 1 & 10 & 12 & 11 & 15 & 4 & 13 \\ 4 & 7 & 12 & 2 & 13 & 16 & 1 & 14 \\ 5 & 8 & 2 & 13 & 15 & 1 & 14 & 11 \\ 6 & 9 & 15 & 3 & 1 & 14 & 11 & 10 \\ 7 & 16 & 13 & 8 & 3 & 11 & 2 & 9 \\ 8 & 14 & 16 & 9 & 6 & 5 & 10 & 2 \\ 9 & 4 & 14 & 16 & 10 & 7 & 3 & 6 \\ 10 & 13 & 4 & 7 & 2 & 12 & 5 & 16 \\ 11 & 15 & 7 & 5 & 14 & 8 & 6 & 3 \\ 12 & 10 & 5 & 1 & 4 & 9 & 13 & 15 \\ 13 & 6 & 8 & 15 & 7 & 3 & 16 & 1 \\ 14 & 11 & 1 & 6 & 8 & 10 & 12 & 7 \\ 15 & 3 & 11 & 10 & 12 & 5 & 9 & 4 \\ 16 & 12 & 6 & 12 & 9 & 2 & 8 & 5 \end{bmatrix}.$$

Golf design was introduced in 1981 by D. F. Robinson [Rob]. $G(3)$ is trivial, but there exists no $G(5)$ [Rob]. W. D. Wallis [Wal] noticed that a golf design of order v is produced from a large set of Steiner triple systems ($LSTS(v)$). Therefore, from Chapter 2, there exists a $G(v)$ for $v \equiv 1,3 \pmod 6$. However, up to 1992, the existence problem for golf design of order v remained open for all $v \equiv 5 \pmod 6$, $v \neq \{5,17\}$. Luc Teirlinck [Teir 9] proved that if a $G(11)$ exists, then a $G(v)$ exists for all $v \equiv 5 \pmod 6$ except 5 and possible 41. But the existence of a $G(11)$ is still in doubt. In [Chang 1], Chang Yanxun, avoiding the existence of $G(11)$, proved that

Theorem 7.1.1. For any odd integer $v > 1$, there exists a $G(v)$ if $v \notin E_0 = \{5,11,23,29,41,53,59,65,77,83,89,113,119,131,155,167,173,179, 185,221, 239,263, 311,353,383,389,401,437,581\}$. And, there exists no $G(5)$.

In this section we will introduce Chang's proof. First, denote $T = \{\text{odd } t : \exists G(3t+2)\}$. Define a kind of special golf design, which is used later. Let $G(v) = \{A_1, \ldots, A_{v-2}\}$ be a golf design of order v based on the set $X = \{x_1, \ldots x_{v-2}\} \bigcup \{a,b\}$ such that $(a,b,x_i), (b,a,x_i), (a,x_i,b), (b,x_i,a), (x_i,a,b)$ and $(x_i,b,a) \in A_i$ for every $i (1 \le i \le v)$. This kind of golf design denoted by $\tilde{G}(v)$. Let $\tilde{M} = \{v : \exists \tilde{G}(v+2)\}$.

7.1.2 49 and 85 $\in T$

Lemma 7.1.2. $\tilde{M} \supseteq \{v : v \equiv 1,5 \pmod 6, v \ge 7\}$.

Proof. Let $v \equiv 1, 5 \pmod 6$ and $v \ge 7$. X is a v−set, $a, b \notin X$ and $a \ne b$. Let

$$LSTS(v+2) = \{(X \bigcup \{a,b\}, \mathcal{B}_x) : x \in X\} \text{ with } \{a,b,x\} \in \mathcal{B}_x.$$

Replacing each unordered triple $\{x,y,z\}$ by six ordered triple : $(x,y,z), (x,z,y), (y,x,z), (y,z,x), (z,x,y), (z,y,x)$, we can obtain a set $\tilde{\mathcal{B}}_x$ of ordered triples from each $\mathcal{B}_x$. Let $J = \{(t,t,t): t \in X \bigcup \{a,b\}\}$, then it is easy to verify that $\{(X \bigcup \{a,b\}, \tilde{\mathcal{B}}_x \bigcup J): x \in X\}$ is a $\tilde{G}(v+2)$. Hence $v \in \tilde{M}$.□

Adapting the notation in this section, the construction 3 in [Teir 9] can be restated as follows.

Lemma 7.1.3. Let K be a set of positive odd integers. If $K \subset \tilde{M}$ and there exists a $B[K,1;v]$, then $v \in \tilde{M}$.

Lemma 7.1.4. $49 \in T$.

Proof. For a prime power q, if a positive integer $t < q^2 - q + 1$, then there exists a $B[\{t, q+t\}, 1; t(q^2+q+1)]$, refer to [Brou 2]. Taking $t = 7$, and $q = 4$ we get a $B[\{7,11\},1;147]$. Therefore, since 7 and $11 \in \tilde{M}$ by Lemma 7.1.2, 147 belongs to $\tilde{M}$ by Lemma 7.1.3. Thus $49 \in T$. □

Let $I_n = \{0, 1, ..., n-1\}$, $a, b \notin I_n$ and $a \neq b$. Suppose there is an

$$LSTS(n+2) = \{(I_n \bigcup \{a,b\}, \mathcal{A}_i) : i \in I_n, \{a,b,i\} \in \mathcal{A}_i\}.$$

We replace each unordered triple in A_i by one of six ordered triple. For example, $\{a,b,i\}$, $\{a,x,y\}$ and $\{b,x,y\}$ $(x,y \in I_n)$ are replaced by (a,b,i), (a,x,y) and (b,x,y) respectively. Denoted by $\tilde{\mathcal{A}}_i$ the obtained set of ordered triples from $\mathcal{A}_i$. If the following condition are satisfied:

(G1) $\{x : x \in I_n, \exists y \in I_n \text{ such that } (a,x,y) \text{ or } (b,x,y) \in \tilde{\mathcal{A}}_i\} = I_n \backslash \{i\}$;

(G2) $\{(x,y) : \exists\ i \text{ such that } (a,x,y) \text{ or } (b,x,y) \in \tilde{\mathcal{A}}_i,\ x,y \in I_n\} = \{(x,y) : x \neq y \in I_n\}$,

then $\{(I_n \bigcup \{a,b\}, \tilde{\mathcal{A}}_i) \colon\ i \in I_n,\ (a,b,i) \in \tilde{\mathcal{A}}_i\}$ is called a *good large set* of Steiner triple systems, denote it by $GLS(n+2)$. In [Lu 1], Lu Jiaxi defined the GLS and gave the following results.

Lemma 7.1.5. There exists a $GLS(p+2)$ for prime power p, $p \equiv 7 \pmod 8$ or $p = 17, 19$.

Now, we give a recursive construction about $G(v)$, using GLS.

Lemma 7.1.6. If both $GLS(n+2)$ and $G(m+2)$ exist, then there exists a $G(mn+2)$.

Costruction. As $m = 1$ is trivial, thus we can suppose $m > 1$. Given

$$GLS(n+2) = \{(I_n \bigcup \{\infty_1, \infty_2\}, \tilde{\mathcal{A}}_i) : i \in I_n, (\infty_1, \infty_2, i) \in \tilde{\mathcal{A}}_i\}$$

$$G(m+2) = \{(I_m \bigcup \{\infty_1, \infty_2\}, \mathcal{B}_j) : j \in I_m\}.$$

Let $(I_m, \circ)$ be a symmetric idempotent quasigroup of order m (note that m is odd by the existence of $G(m+2)$). Let $\alpha = (0, 1, ..., m-1)$ be a cycle of order m. Construct mn set $\mathcal{T}_{ij}$ $(i \in I_n, j \in I_m)$ of ordered triples on the set $S = (I_n \times I_m) \bigcup \{\infty_1, \infty_2\}$. Each $\mathcal{T}_{ij}$ contains seven parts as follows:

(1) $(s, s, s), s \in S$.

(2) $((x,u),(y,v),(z,(u\circ v)\alpha^j)), ((x,u),(z,(u\circ v)\alpha^j),(y,v)), ((y,v),(x,u),(z,(u\circ v)\alpha^j)), ((y,v),(z,(u\circ v)\alpha^j),(x,u)), ((z,(u\circ v)\alpha^j),(x,u),(y,v)), ((z,(u\circ v)\alpha^j),(y,v),(x,u))$, where $(x,y,z) \in \tilde{\mathcal{A}}_i$, $x,y,z,\in I_n$ and $u,v \in I_m$.

(3) $((x,u),(x,v),(y,(u\circ v)\alpha^j)), ((x,u),(y,(u\circ v)\alpha^j),(x,v)), ((y,(u\circ v)\alpha^j),(x,u),(x,v))$, where (∞_1,x,y) or $(\infty_2,x,y) \in \tilde{\mathcal{A}}_i$, $x,y \in I_n$ and $u \neq v \in I_m$.

(4) $(\infty_1,(x,u),(y,u\alpha^j))$, $((x,u),\infty_1,(y,u\alpha^j))$, $(\infty_2,(y,u\alpha^j),(x,u))$, $((y,u\alpha^j),\infty_2,(x,u))$, where (∞_1,x,y) or $(\infty_2,x,y) \in \tilde{\mathcal{A}}_i$, $x,y \in I_n$ and $u \in I_m$.

(5) $((x,u),(y,u\alpha^j),\infty_1), ((y,u\alpha^j),(x,u),\infty_1)$, where $(\infty_1,x,y) \in \tilde{\mathcal{A}}_i$, $x,y \in I_n$ and $u \in I_m$.

(6) $((x,u),(y,u\alpha^j),\infty_2), ((y,u\alpha^j),(x,u),\infty_2)$, where $(\infty_2,x,y) \in \tilde{\mathcal{A}}_i$, $x,y \in I_n$ and $u \in I_m$.

(7) $((i,u),(i,v),(i,w))$, where $(u,v,w) \in \mathcal{B}_j \backslash \{(t,t,t) : t \in I_m \bigcup \{\infty_1, \infty_2\}\}$ (whenever ∞_1 or ∞_2 appears for u,v,w, omit the first coordinate i).

Proof. 1° Each $\mathcal{T}_{ij}$ is a Latin square of order $mn+2$ (i.e. an $OA(mn+2)$). Direct calculation shows that

$$|\mathcal{T}_{ij}| = (mn+2)+m^2(n-1)(n-2)+3m(m-1)(n-1)+4m(n-1)+m(n-1)$$
$$+m(n-1)+(m+2)(m+1) = (mn+2)^2.$$

Here, we only prove the fact that "for any ordered pair $P = (\alpha, \beta)$, $\alpha, \beta \in S$ there exists $\gamma \in S$ such that $(\alpha,\beta,\gamma) \in \mathcal{T}_{ij}$", other case is similar. All the possibilities are exhausted as follows.

(i) $P = (s,s), s \in S$, is contained in (1) of $\mathcal{T}_{ij}$.

(ii) $P = (\infty_1, \infty_2)$. There exists $w \in I_m$ such that $(\infty_1, \infty_2, w) \in \mathcal{B}_j$, then P is contained in (7) of $\mathcal{T}_{ij}$. ($P = (\infty_2, \infty_1)$ is similar.)

(iii) $P = (\infty_1,(x,u)), x \in I_n, u \in I_m$. If $x = i$, there exists $(\infty_1,u,v) \in \mathcal{B}_j$, P is contained in (7) of $\mathcal{T}_{ij}$. If $x \in I_n \backslash \{i\}$, there exists $y \in I_n$ such that (∞_1, x, y) or $(\infty_2, x, y) \in \tilde{\mathcal{A}}_i$ by $(G1)$, then P is contained in (4) of $\mathcal{T}_{ij}$. (It is similar for $P = ((x,u), \infty_1)$, $(\infty_2, (x,u))$ and $((x,u), \infty_2)$).

(iv) $P = ((x,u),(x,v)), x \in I_n, u \neq v \in I_m$. If $x = i$, P is contained in (7). If $x \in I_n \setminus \{i\}$, by (G1), there exists $y \in I_n$ such that (∞_1, x, y) or $(\infty_2, x, y) \in \tilde{A}_i$, then P is contained in (3).

(v) $P = ((x,u),(y,v))$, $x \neq y \in I_n$, $u, v \in I_m$. There exists $z \in I_n \bigcup \{\infty_1, \infty_2\}$ such that $\{x,y,z\} \in \mathcal{A}_i$. If $z \in I_n$, without loss the generality, let $(x,y,z) \in \tilde{\mathcal{A}}_i$, then P is contained in (2). If $z = \infty_1$, (∞_1, x, y) or $(\infty_1, y, x) \in \tilde{\mathcal{A}}_i$, P is contained in (5) when $u\alpha^j = v$, otherwise P in (3). It is similar for $z = \infty_2$.

2° It is easy to see that each $\mathcal{T}_{ij}$ is symmetric and idempotent.

3° $\{(S, \mathcal{T}_{ij}) : i \in I_n, j \in I_m\}$ is a golf design of order $mn + 2$.

Let $J = \{(s,s,s) : s \in S\}$. Then every triple of $\mathcal{T}_{ij} \setminus J$ have three distinct compotent. We only need to verify that any ordered triple of distinct elements from S is contained in one part of (2)—(7) of some $\mathcal{T}_{ij}$.

(i) $T = ((i,u),(i,v),(i,w)), u, v, w \in I_m \bigcup \{\infty_1, \infty_2\}$, where u, v, w are distinct. There exists $j \in I_m$ such that $(u,v,w) \in \mathcal{B}_j$, then T is contained in (7) of $\mathcal{T}_{ij}$.

(ii) $T = (\infty_1, (x,u), (y,v))$, where $x \neq y \in I_n$, $u, v \in I_m$. From the condition (G2), there exist $i \in I_n$ and $j \in I_m$ such that (∞_1, x, y) or $(\infty_2, y, x) \in \tilde{\mathcal{A}}_i$ and $u\alpha^j = v$. Then T is contained in (4) of $\mathcal{T}_{ij}$ (It is similar for $T = ((x,u), \infty_1, (y,v)), (\infty_2, (x,u), (y,v)), ((x,u), \infty_2, (y,v))$.

(iii) $T = ((x,u),(y,v),\infty_1)$, where $x \neq y \in I_n$ and $u, v \in I_m$. There exists $i \in I_n$ such that (∞_1, x, y) or $(\infty_1, y, x) \in \tilde{\mathcal{A}}_i$ (since $\{\mathcal{A}_i\}_{i \in I_n}$ is an $LSTS$). And there exists $j \in I_m$ such that $u\alpha^j = v$ or $v\alpha^j = u$. Then T is contained in (5) of $\mathcal{T}_{ij}$ (similarly, $T = ((x,u),(y,v),\infty_2)$ is contained in part (6)).

(iv) $T = ((x,u),(x,v),(y,w))$, where $x \neq y \in I_m$, $u, v, w \in I_m$ and $u \neq v$. From $(G2)$, there exists $i \in I_n$ such that (∞_1, x, y) or $(\infty_2, y, x) \in \tilde{A}_i$. And there exists $j \in I_m$ such that $(u \circ v)\alpha^j = w$. Then T is contained in part (3) (It is similar for $T = ((x,u),(y,w),(x,v)), ((y,w),(x,u),(x,v))$).

(v) $T = ((x,u),(y,v),(z,w))$, where x, y, z are distinct elements from I_n and $u, v, w \in I_m$. There exists $i \in I_n$ such that $\{x,y,z\} \in \mathcal{A}_i$ (without loss the generality $(x, y, z) \in \tilde{\mathcal{A}}_i$). there exists $j \in I_m$ such that $(u \circ v)\alpha^j = w$. Then T is contained in (2). □

Corollary 7.1.7. $85 \in T$.

Proof. By Lemma 7.1.5 there exists a $GLS(19)$. As well as we have a $G(15+2)$ by Example 2. Applying Lemma 7.1.6 (taking $m = 15$ and $n = 17$), we can obtain a $G(15 \times 17 + 2)$. Thus $5 \times 17 = 85$ belongs to T. □

7.1.3 The range of set T

Lemma 7.1.8. If $t \in T$, then $2t + 1 \in T$.

Proof. There exists a $G(3t+2)$ by $t \in T$. Using Rosa's construction $LSTS(v) \longrightarrow LSTS(2v+1)$ (refer to [Ro 1]), we can give a recursive construction $G(3t+2) \longrightarrow G(6t+5)$. Thus, $2t + 1 \in T$. □

A *group-divisible design* $GD[K, \lambda, M; v]$ is a triple $(X, \mathcal{G}, \mathcal{A})$, where X is a v−set, $\mathcal{G}$ is a partition of X into m_i-subsets (called *groups*, $m_i \in M$) and $\mathcal{A}$ is a collection of k_i-subsets of X (called *blocks*, $k_i \in K$) such that $| B \cap G | \leq 1$ for all $B \in \mathcal{A}$ and $G \in \mathcal{G}$, and such that any 2-subset P of X with $| P \bigcap G | \leq 1$ for all $G \in \mathcal{G}$ is contained in exactly λ blocks of $\mathcal{A}$. The *type* of a $GD[K, \lambda, M; v]$ is $m_1{}^{r_1} m_2{}^{r_2} \cdots m_t{}^{r_t}$, if $M = \{m_1, m_2, \cdots, m_t\}$ and there are r_i groups with size m_i in $\mathcal{G}$, $1 \leq i \leq t$. A $GD[3, 1, M; v]$ is called a 3-GDD and a 3-GDD of type T is denoted by 3-$GDD(T)$.

Lemma 7.1.9. If there is a 3-GDD, $GD[K, 1, M; v]$, and $K \in T$, then $v \in T$.

Proof. Adapting the notation in the section, the proposition 3 in [Teir 9] can be restated as this Lemma. □

Corollary 7.1.10. If $n \in T$ then $3n \in T$.

Proof. For any integer $n > 1$, there exists a $TD(3, n)$, i.e. a 3-$GDD(n^3)$. Therefore $3n \in T$ by $n \in T$ and Lemma 7.1.9. □

Lemma 7.1.11. [ColHoRe] For nonnegative integers g, t, u, $g \geq 1$, $t \geq 3$ and $u \leq g(t-1)$, if $g^2t(t-1)/2 + gtu \equiv 0 \pmod 3$, $g(t-1) + u \equiv 0 \pmod 2$, and gt is even (when $u \neq 0$), then there exists a 3-$GDD(g^t u^1)$.

Remark: For positive integers g, t, u, if $6|t$, $u \leq g(t-1)$ and $g, u \in T$, then all conditions of Lemma 7.1.11 are satisfied (note that g must be odd here). Furthermore we can get $gt+u \in T$ by Lemma 7.1.9.

Corollary 7.1.12. If $s \geq 1, u \in T$ and $u \leq 5(6s-1)$, then $30s+u \in T$.

Proof. Take $g=5$, $t=6s$. As $5 \in T$ (there exists $G(17)$, it is easy to see that the conclusion holds by Remark mentioned above. □

Corollary 7.1.13. If $s \geq 1, u \in T$ and $u \leq 11(6s-1)$, then $66s+u \in T$.

Proof. Take $g=11, t=6s$. As $11 \in T$ ($5 \in T \rightarrow 11 \in T$ by Lemma 7.1.8), it is easy to see that the conclusion holds by Remark above. □

In the next Lemma, we will use Wilson's Fundamental Construction (refer to [Wil 2]):

Suppose that $(X, \mathcal{G}, \mathcal{A})$ is a GDD, and let $w: X \rightarrow N_0$ be a weighting, where N_0 is the set of all nonnegative integers. For every $x \in X$, let S_x be a set of $w(x)$ "copies" of x. For every $A \in \mathcal{A}$, if $(\bigcup_{x \in A} S_x, \{S_x : x \in A\}, \mathcal{B}_A)$ is a GDD, then $(\bigcup_{x \in X} S_x, \{\bigcup_{x \in G} S_x : G \in \mathcal{G}\}, \bigcup_{A \in \mathcal{A}} \mathcal{B}_A)$ is also a GDD.

Lemma 7.1.14. For integers t, n, $0 \leq t \leq n$, if there exists a $TD(5,n)$, then there exists a 3- $GDD(n^3(3n)^1(n+2t)^1)$.

Proof. Let $(X, \{G_0, G_1, G_2, G_3, G_4\}, \mathcal{A})$ be a $TD(5,n)$, where $G_0 = \{y_1, \ldots, y_t, \ldots, y_n\}$. Define a weighting function $w: X \rightarrow N_0$ such that

$$w(x) = \begin{cases} 1, & x \in G_2 \bigcup G_3 \bigcup G_4 \\ 3, & x \in G_1 \text{ and } x \in \{y_1, .., y_t\} \subset G_0 \\ 0, & \text{else.} \end{cases}$$

By Fundamental Construction, the desired 3-GDD with group type $1^4 3^1$ and $1^3 3^2$ exists from the existence of $B[3,1;7]$ and $RB[3,1;9]$. The conclusion follows. □

Corollary 7.1.15. If $n, n+2t \in T$, $n > 3$ and $0 \leq t \leq n$, then $7n+2t \in T$.

Proof. Since $n > 3$ and $n \in T$, thus $n \notin \{2,3,6,10\}$ and there exists a $TD(5,n)$. By Lemma 7.1.14, there exists a $3-GDD(n^3(3n)^1(n+2t)^1)$. As well as $3n \in T$ from $n \in T$ by Corollary 7.1.10. Therefore, applying Lemma 7.1.9 we obtained $7n+2t \in T$. □

Lamma 7.1.16. 5, 11, 15, 23, 31, 33, 47, 49, 53, 63, 67, 69, 75, 77, 81, 85, 89, 91, 97, 99, 105, 111, 115, 119, 141, 147, 149, 163, 177, 179, 207, 223, 237, 253, 267, 283, 297, 313 $\in T$.

Proof. Firstly, $5, 11, 49, 85 \in T$, by Lemma 7.1.4, Corollary 7.1.7 and the proofs of Corollary 7.1.12 and Corollary 7.1.13. Furthermore,

$11 \in T \to 23 \in T \to 47 \in T$ by Lemma 7.1.8;

$5, 11, 23, 47, 49 \in T \to 15, 33, 69, 141, 147 \in T$ by Lemma 7.1.10;

$15, 33 \in T \to 31, 67 \in T$ and $31, 81 \in T \to 63, 163 \in T$ by Lemma 7.1.8.
Using Lemma 7.1.12 and Lemma 7.1.15 respectively, some orders $30s+u \in T$ and $7 \times 11 + 2t \in T$ can be obtained as follows:

s	1	2	2	3
u	23	15	31	15
$30s + u$	53	75	91	105

t	0	2	6	10	11
$11 + 2t$	11	15	23	31	33
$7 \times 11 + 2t$	77	81	87	97	99

And, using Lemma 7.1.13, some orders $66s + u \in T$ can be obtained in the following table (where $45 \in T$ is from $15 \in T$ by Lemma 7.1.10).

s	1	1	1	2	2	2	2	2	3	3	4	4
u	45	49	53	45	47	75	91	105	69	85	33	49
$66s + u$	111	115	119	177	179	207	223	237	267	283	297	313

Lastly, using the Remark (after Lemma 7.1.11) and taking $t = 6$, $g = 23$, we can get $149 \in T$ (for $u = 11$) and $253 \in T$ (for $u = 115$). □

Lemma 7.1.17. For any odd integer $t > 1$, if $t \notin E$ then $t \in T$, where $E = \{3, 7, 9, 13, 17, 19, 21, 25, 27, 29, 37, 39, 43, 51, 55, 57, 59, 61, 73, 79, 87, 103, 117, 127, 129, 133, 145, 193\}$.

Proof. Applying Lemma 7.1.12, we have the fact:

"If $u \in T$ and $s > s_0 = \lfloor \frac{u+5}{30} \rfloor + 1$, then $30s + u \in T$."

$t \pmod{30}$	u	s_0	known orders $30i+u$ $(0<i<s_0)$	possible exception
1	31	2		61
3	33	2	63	3
5	5	1		
7	67	3	97	7, 37, 127
9	69	3	99	9,39,129
11	11	1		
13	163	6	223,253,283,313	13,43,73,103,133,193
15	15	1		
17	47	2	77	17
19	49	2		19,79
21	81	3	111,141	21,51
23	23	1		
25	85	3	115	25,55,145
27	147	6	177,207,237,267,297	27,57,87,117
29	89	4	119,149,179	29, 59

In this table, by the fact mentioned above, we consider the all cases of odd integers modulo 30, where these orders listing in the second and the fourth columns are from Lemma 7.1.16. □

7.1.4 The conclusion

The proof of Theorem 7.1.1.

If $v \equiv 1, 3 \pmod 6$ and $v \neq 7$, there exists an $LSTS(v)$, which implies a $G(v)$. The existence of $G(7)$ comes from [Rob]. If $v \equiv 5 \pmod 6$ and $v \notin E_0$, let $v = 3t + 2$, then t must be odd and $t \notin E$ (since there exists no G(5), thus $t = 1 \notin T$). Thereby, by Lemma 7.1.17, $t \in T$ when $t \notin E$. Hence there exists a $G(3t + 2) = G(v)$. □

7.2 About large sets of *GDD*

7.2.1 *GDD* and $LS(t^u)$

In this section, some large sets of GDD will be discussed. In what follows, we will consider only such GDD that with block size 3 and index 1, i.e. $GD[3, 1, M; v]$ or 3-GDD. A 3-GDD of type T is denoted by 3-$GDD(T)$. Two 3-GDDs are called *disjoint* if they have the same group set and their block sets are disjoint. A set of more than two 3-GDDs (having the same group set) are called disjoint if each pair of them are disjoint. It is easy to see that there are $\frac{u(u-1)t^2}{6}$ blocks in a 3-$GDD(t^u)$, and there exists a 3-$GDD(t^u)$ only if

$$u(u-1)t^2 \equiv 0 \pmod 6 \quad \text{and} \quad (u-1)t \equiv 0 \pmod 2 \qquad (*)$$

It is well known that these conditions are also sufficient (refer to [BJL]). Furthermore, since there are $\frac{u(u-1)(u-2)}{6}t^3$ triples, whose elements belong to different groups, the maximum number of disjoint 3-GDDs of type t^u is $t(u-2)$. Similarly, for $s \geq t$, we can see that the maximum number of disjoint 3-GDDs of type $t^u s^1$ is $t(u-1)$. Such a collection of disjoint 3-GDDs will be called a *large set* of GDD's, and denoted by $LS(t^u)$ and $LS(t^u s^1)$, respectively. Below, we will discuss the 3-GDD with type t^u and their large set.

7.2.2 Examples and known results

By condition $(*)$, the necessary condition that there exists an $LS(t^u)$ is one of the following cases:

(LG1) $u \equiv 1, 3 \pmod 6$ for any t;

(LG2) $u \equiv 0, 4 \pmod 6$ and $t \equiv 0 \pmod 2$;

(LG3) $u \equiv 5 \pmod 6$ and $t \equiv 0 \pmod 3$;

(LG4) $u \equiv 2 \pmod 6$ and $t \equiv 0 \pmod 6$.

A 3-$GDD(1^u)$ is just a *Steiner triple system* of order u, $STS(u)$. Therefore, $LS(1^u)$ is just an $LSTS(u)$. By Theorem 2.4.1, the existence spectrum of $LSTS$ has been completed. Thus we have

Theorem 7.2.1. There exists an $LS(1^u)$ if and only if $u \equiv 1$ or $3 \pmod 6$ and $u \neq 7$.

$LS(t^u)$ is an interesting and natural generalization of large set of Steiner triple system, and it can be applied in cryptographic to construct perfect threshold schemes. As examples, let us exhibit some $LS(t^u)$ for some small orders.

Example 1. An $LS(2^4) = \{(X, \mathcal{G}, \mathcal{B}_i) : 1 \leq i \leq 4\}$, where $X = \{1,2,3,4,5,6,7,8\}$, $\mathcal{G} = \{\{1,2\},\{3,4\},\{5,6\},\{7,8\}\}$, and $\mathcal{B}_i$ is listed as follows:

$\mathcal{B}_1$	$\mathcal{B}_2$	$\mathcal{B}_3$	$\mathcal{B}_4$
$\{1,3,5\}$	$\{1,3,6\}$	$\{1,3,7\}$	$\{1,3,8\}$
$\{1,4,8\}$	$\{1,4,7\}$	$\{1,4,5\}$	$\{1,4,6\}$
$\{1,6,7\}$	$\{1,5,8\}$	$\{1,6,8\}$	$\{1,5,7\}$
$\{2,3,7\}$	$\{2,3,8\}$	$\{2,3,6\}$	$\{2,3,5\}$
$\{2,4,6\}$	$\{2,4,5\}$	$\{2,4,8\}$	$\{2,4,7\}$
$\{2,5,8\}$	$\{2,6,7\}$	$\{2,5,7\}$	$\{2,6,8\}$
$\{3,6,8\}$	$\{3,5,7\}$	$\{3,5,8\}$	$\{3,6,7\}$
$\{4,5,7\}$	$\{4,6,8\}$	$\{4,6,7\}$	$\{4,5,8\}$

Example 2. An $LS(3^5) = \{(X, \mathcal{G}, \mathcal{A}_i) : \quad i \in Z_9\}$, where $X = \{\infty_j : \quad j \in Z_6\} \bigcup Z_9$, $\mathcal{G} = \{\{\infty_0, \infty_2, \infty_4\}, \{\infty_1, \infty_3, \infty_5\}, \{0,3,6\}, \{1,4,7\}, \{2,5,8\}\}$ and $\mathcal{A}_0$ contains the following blocks:

$\{0,1,8\}$	$\{2,4,6\}$	$\{3,4,8\}$
$\{\infty_0,\infty_1,6\}$	$\{\infty_0,\infty_3,3\}$	$\{\infty_0,\infty_5,8\}$
$\{\infty_2,\infty_1,8\}$	$\{\infty_2,\infty_3,1\}$	$\{\infty_2,\infty_5,4\}$
$\{\infty_4,\infty_1,4\}$	$\{\infty_4,\infty_3,0\}$	$\{\infty_4,\infty_5,2\}$
$\{\infty_0,1,2\}$	$\{\infty_0,5,7\}$	$\{\infty_0,0,4\}$
$\{\infty_1,2,3\}$	$\{\infty_1,0,7\}$	$\{\infty_1,1,5\}$
$\{\infty_2,5,6\}$	$\{\infty_2,0,2\}$	$\{\infty_2,3,7\}$
$\{\infty_3,4,5\}$	$\{\infty_3,6,8\}$	$\{\infty_3,2,7\}$
$\{\infty_4,7,8\}$	$\{\infty_4,3,5\}$	$\{\infty_4,1,6\}$
$\{\infty_5,6,7\}$	$\{\infty_5,1,3\}$	$\{\infty_5,0,5\}$

Furthermore, $\mathcal{A}_i = \mathcal{A}_0 + i = \{\{a+i, b+i, c+i\} : \{a,b,c\} \in \mathcal{A}_0\}$, where $i \in Z_9$ and $\infty_j + i = \infty_j$ for any $j \in Z_6$.

Using computer, the author has obtained (see [LeiKang]) another $LS(3^5) = \{(I_5 \times Z_3, \mathcal{G}, \mathcal{B}_j) : 1 \leq j \leq 9\}$, where $\mathcal{G} = \{\{x\} \times Z_3\}$ and each $\mathcal{B}_j$ consists of ten parts as follows.

(x,y,z)	$\mathcal{B}_1$	$\mathcal{B}_2$	$\mathcal{B}_3$	$\mathcal{B}_4$	$\mathcal{B}_5$	$\mathcal{B}_6$	$\mathcal{B}_7$	$\mathcal{B}_8$	$\mathcal{B}_9$
012	00	01	02	12	10	11	21	22	20
013	10	20	12	21	02	22	01	11	00
014	20	10	22	00	21	01	11	02	12
023	20	00	01	10	21	02	12	22	11
024	10	20	11	22	02	12	21	00	01
034	00	12	21	20	10	02	22	01	11
123	11	22	12	02	10	01	21	00	20
124	22	01	00	11	21	20	02	12	10
134	11	20	10	01	02	00	21	22	12
234	22	11	02	12	21	01	00	10	20

When the entry st appears in the intersection of the row (x,y,z) and the column $\mathcal{B}_j$, the system $\mathcal{B}_j$ contains the following three triples:

$$\{(x,i),(y,i+s),(z,i+s+t)\},\ i\in Z_3.$$

For example, the system $\mathcal{B}_1$ consists of these triples:

$\{(0,i),(1,i),(2,i)\},\{(0,i),(1,i+1),(3,i+1)\},$
$\{(0,i),(1,i-1),(4,i-1)\},\{(0,i),(2,i-1),(3,i-1)\},$
$\{(0,i),(2,i+1),(4,i+1)\},\{(0,i),(3,i),(4,i)\},$
$\{(1,i),(2,i+1),(3,i-1)\},\{(1,i),(2,i-1),(4,i+1)\},$
$\{(1,i),(3,i+1),(4,i-1)\},\{(2,i),(3,i-1),(4,i+1)\}.$

It is easy to see that two $LS(3^5)$ above are nonisomorphic, since all $\mathcal{A}_i$ are mutually isomorphic but the second structure have not the property.

Example 3. An $LS(2^7)=\{((Z_5\bigcup\{\infty_1,\infty_2\})\times Z_2,\ \mathcal{G},\mathcal{A}_{ti}):\ t\in Z_5,\ i\in Z_2\}$, where $\mathcal{G}=\{\{u\}\times Z_2:\ u\in Z_5\bigcup\{\infty_1,\infty_2\}\}$, and $\mathcal{A}_{ti}$ contains the following blocks (where $\{x,y,z\}\in\{\{\infty_1,\infty_2,t\},\ \{t,1+t,4+t\},\ \{t,2+t,3+t\}\}$ and $j\in Z_2$):

(i) $\{(x,i),(y,i),(z,i)\},$
$\{(x,i),(y,i+1),(z,i+1)\},$
$\{(x,i+1),(y,i),(z,i+1)\},$
$\{(x,i+1),(y,i+1),(z,i)\},$

(ii) $\{(\infty_1,i),(3+t,i+j),(4+t,i+1+j)\},$
$\{(\infty_1,i),(2+t,i+j),(1+t,i+j)\},$
$\{(\infty_1,i+1),(1+t,i+j),(3+t,i+j)\},$

$\{(\infty_1, i+1), (4+t, i+j), (2+t, i+1+j)\}$,
$\{(\infty_2, i), (3+t, i+j), (4+t, i+j)\}$,
$\{(\infty_2, i), (2+t, i+j), (1+t, i+j+1)\}$,
$\{(\infty_2, i+1), (1+t, i+j), (3+t, i+j+1)\}$,
$\{(\infty_2, i+1), (4+t, i+j), (2+t, i+j)\}$.

The papers [ChenSt 1-3] and [ChenLinSt] presented several constructing methods and some $LS(t^u)$ for other small orders, such as $LS(2^6)$ and $LS(3^7)$. They are summarized in the following.

Theorem 7.2.2. (1) If there exists an $LS(t^u)$, then there exists an $LS((ts)^u)$ for any integer $s \geq 2$.

(2) For even t, if there exists an $LS(t^{u+2})$, then there exists an $LS(t^{uv+2})$ for any $v \equiv 1$ or 5 (mod 6).

(3) If there exists an $LS(t^u)$, $u \neq 6$, then there exists an $LS(t^{3u})$.

(4) If there exist both an $LS(t^3s^1)$ and an $LS(t^us^1)$, $t \leq s$, $u \neq 2, 6$, then there exists an $LS(t^{3u}s^1)$.

(5) If there exists an $LS(t^{u-1})$, u even, $u \geq 6$ and $t = 3$ or 6, then there exists an $LS(t^{2u-1})$.

(6) If there exist both a $GLS(p+2)$ and an $LS(t^{u+2})$, then there exists an $LS(t^{uv+2})$.

Here, $GLS(p+2)$ was introduced by Lu Jiaxi in [Lu 2], the definition of which has been introduced in the last section.

After Theorem 7.2.1 and 7.2.2, the existence of $LS(t^u)$ is still open. For $u \equiv 1$ or 3 (mod 6), $u > 7$ and $t \geq 1$, there exists an $LS(t^u)$ by Theorem 7.2.1 and Theorem 7.2.2 (1). There exists no $LS(1^7)$, but there exist both an $LS(2^7)$ and an $LS(3^7)$. So an $LS((pt)^7)$ exists for $p = 2$ or 3 and any t. Thereby, the case (LG1) will be sufficient if there exists $LS(p^7)$ for prime $p > 3$. Lei Jianguo and Chen Demeng have completed this work, which will be introduced in next subsection.

7.2.3 The construction of $LS(p^7)$

Let $X = \{\infty_1, \infty_2\} \bigcup Z_5$. Suppose p is prime, $p \geq 5$. We try to construct an $LS(p^7)$ on set $X \times Z_p$. For the distinct elements $x, y, z \in X$ and $n \in Z_p$, we define a triple system, $\{x, y, z\}_{-n}$, which consists of the following p^2 triples:

$$\{\{(x,i),(y,j),(z,k)\}:\ i+j+k\equiv 3n \pmod p\}.$$

Note that $(Y,\ \mathcal{G}_Y,\ \{x,y,z\}\_n)$ is a 3-$GDD(p^3)$, where element set $Y = \{x,y,z\}\times Z_p$, group set $\mathcal{G}_Y$ consists of three groups $\{x\}\times Z_p, \{y\}\times Z_p$ and $\{z\}\times Z_p$. Now construct $5p$ block sets $\mathcal{B}_{mn}$ $(m\in Z_5,\ n\in Z_p)$ on the set $X\times Z_p$ as follows:

(1) $\mathcal{B}_{m0}$: $\{m,1+m,4+m\}\_0$, $\{m,2+m,3+m\}\_0$, $\{\infty_1,4+m,2+m\}\_0$, $\{\infty_1,3+m,1+m\}\_1$, $\{\infty_2,3+m,4+m\}\_0$, $\{\infty_2,1+m,2+m\}\_1$, $\{\infty_1,\infty_2,m)\}\_0$;

(2) $\mathcal{B}_{m1}$: $\{m,1+m,3+m\}\_1$, $\{m,2+m,4+m\}\_-1$, $\{\infty_1,3+m,4+m\}\_0$, $\{\infty_1,1+m,2+m\}\_1$, $\{\infty_2,4+m,1+m\}\_0$, $\{\infty_2,2+m,3+m\}\_-1$, $\{\infty_1,\infty_2,m\}\_1$;

(3) $\mathcal{B}_{mn}$: $(2\le n\le \frac{p-1}{2})$ $\{m,1+m,4+m\}\_n$, $\{m,2+m,3+m\}\_n$, $\{\infty_1,1+m,3+m\}\_n$, $\{\infty_1,2+m,4+m\}\_-n$, $\{\infty_2,3+m,4+m\}\_n$, $\{\infty_2,1+m,2+m\}\_-n$, $\{\infty_1,\infty_2,m\}\_n$;

(4) $\mathcal{B}_{mn}$: $(\frac{p+1}{2}\le n\le p-2)$ $\{m,1+m,4+m\}\_n$, $\{m,2+m,3+m\}\_n$, $\{\infty_1,3+m,4+m\}\_n$, $\{\infty_1,1+m,2+m\}\_-n$, $\{\infty_2,2+m,4+m\}\_-n$, $\{\infty_2,1+m,3+m\}\_n$, $\{\infty_1,\infty_2,m\}\_n$;

(5) $\mathcal{B}_{m,p-1}$: $\{m,3+m,4+m\}\_1$, $\{m,1+m,2+m\}\_-1$, $\{\infty_1,1+m,4+m\}\_-1$, $\{\infty_1,2+m,3+m\}\_-1$, $\{\infty_2,2+m,4+m\}\_1$, $\{\infty_2,1+m,3+m\}\_-1$, $\{\infty_1,\infty_2,m\}\_-1$.

It is not hard to verify that each $(X\times Z^p, \{\{x\}\times Z_p;\ x\in X\}, \mathcal{B}_{mn})$, $m\in Z_5,\ n\in Z_p$, is a 3-$GDD(p^7)$ and the $5p$ systems are disjoint. Thus, an $LS(p^7)$ is obtained. Furthermore, we have

Theorem 7.2.3. There exists an $LS(t^u)$ for $u\equiv 1$ or $3 \pmod 6$ and any t, except $(t,u)=(1,7)$.

Proof. When $u\ne 7$, the conclusion holds by Theorem 7.2.1 and Theorem 7.2.2(1). When $u=7$ and $t>1$, let p be the smallest prime divisor of t and denote $t=ps$. There exists an $LS(p^7)$ by the construction above (when $p\ge 5$) and the existence of $LS(2^7)$ and $LS(3^7)$. Furthermore, by Theorem 7.2.2(1), there exists an $LS(t^7)$ for any $t>1$. Therefore, for $u\equiv 1$ or $3 \pmod 6$ and any integer t, there exists an $LS(t^u)$ except $LS(1^7)$. □

7.2.4 A recursive construction and recent results

The construction in Theorem 7.2.2 is weakly for constructing an $LS(t^u)$ satisfying condition (LG3). Chen Demeng and Lei Jianguo gave a construction, which will be used to construct some $LS(t^u)$ that satisfies (LG3).

Lemma 7.2.4. For odd v, if there exists an $LS(t^{v+2})$, then there exists an $LS(t^{5v+2})$.

Construction. Suppose v and t are odd, $t \geq 3$ and X is a v-set. Let $(Z_t, \circ)$ be a symmetric idempotent quasigroup and α is a cyclic permutation with length vt on the set $X \times Z_t$.

1° For any $a, b \in Z_5$ and $u \in \{\infty_1, \infty_2\}$, define

$Y_{abu} = ((\{a,b\} \times X) \bigcup \{u\}) \times Z_t$,

$\mathcal{G}_{abu} = \{(a,x) \times Z_t : \ x \in X\} \bigcup \{\{u\} \times Z_t\} \bigcup \{\{b\} \times X \times Z_t\}$,

$B(u,a,b,k) = C(u,a,b,k) \bigcup S(u,a,b,k)$,

where

$C(u,a,b,k) = \{(a,(x, j \circ m)), (a,(y,j)), (b, \alpha^k(\frac{x+y}{2}, m))\}$ with $x \neq y \in X$ and $j, m \in Z_t$;

$S(u,a,b,k) = \{(u, j \circ m), (a,(x,j)), (b, \alpha^k(x,m))\}$ with $x \in X$ and $j, m \in Z_t$.

Then it is not difficult to verify that

$$LS(t^{v+1}(tv)^1) = \{(Y_{abu}, \mathcal{G}_{abu}, B(u,a,b,k)) : \ k \in Z_{tv-1}\}$$

is an $LS(t^{v+1}(tv)')$.

Remark: Since $(Z_t, \circ)$ is symmetric, thus $j \circ m = m \circ j$ for any $j, m \in Z_t$. Thereby,

$$\begin{aligned} S(u,a,b,k) &= \{\{(u, j \circ m), (a,(x,j)), (b, \alpha^k(x,m))\} : x \in X, j, m \in Z_t\} \\ &= \{\{(u, m \circ j), (b, \alpha^k(x,m)), (a,(x,j))\} : x \in X, j, m \in Z_t\} \\ &= \{\{(u, j \circ m), (b, \alpha^k(x,j)), (a,(x,m))\} : x \in X, j, m \in Z_t\} \\ &= \{\{(u, j \circ m), (b,(x,j)), (a, \alpha^{-k}(x,m)\} : x \in X, j, m \in Z_t\} \\ &= S(u,b,a,-k). \end{aligned}$$

2° For any triple $\{a,b,c\}$ from Z_5, by Theorem 7.2.3, there exists an

$$LS((tv)^3) = \{(X_{abc}, \mathcal{G}_{abc}, D(a,b,c,k)) : \ k \in Z_{tv}\},$$

where $X_{abc} = \{a,b,c\} \times X \times Z_t$ and $\mathcal{G}_{abc} = \{\{z\} \times X \times Z_t\colon\ z \in \{a,b,c\}\}$.

3° For any $a \in Z_5$, suppose there exists

$$LS(t^{v+2}) = \{(X_a, \mathcal{G}_a, T(\infty_1, \infty_2, a, k));\ k \in Z_{tv}\},$$

where $X_a = ((\{a\} \times X) \bigcup \{\infty_1, \infty_2\}) \times Z_t$ and $\mathcal{G}_a = \{\{(a,x)\} \times Z_t :\ x \in X\} \bigcup \{\{\infty_j\} \times Z_t :\ j = 1,2\}$.

4° Construct an

$$LS(t^{5n+2}) = \{(Y \times Z_t, \mathcal{G}, \mathcal{B}_k^a) :\ k \in Z_{tv}, a \in Z_5\},$$

where $Y = (Z_5 \times X) \bigcup \{\infty_1, \infty_2\}$, $\mathcal{G} = \{\{y\} \times Z_t\colon\ y \in Y\}$. And, each block set B_k^a consists of the following parts:

(1) $\mathcal{B}_0^a$: $D(a, 1+a, 4+a, 0)$, $D(a, 2+a, 3+a, 0)$, $B(\infty_1, 1+a, 3+a, 1)$, $B(\infty_1, 4+a, 2+a, 2)$, $B(\infty_2, 3+a, 4+a, 2)$, $B(\infty_2, 2+a, 1+a, 0)$, $T(\infty_1, \infty_2, a, 0)$.

(2) $\mathcal{B}_1^a$: $D(a, 1+a, 3+a, 1)$, $D(a, 2+a, 4+a, -1)$, $B(\infty_1, 1+a, 4+a, -2)$, $B(\infty_1, 3+a, 2+a, 2)$, $B(\infty_2, 4+a, 3+a, -1)$, $B(\infty_2, 2+a, 1+a, 1)$, $T(\infty_1, \infty_2, a, 1)$.

(3) $\mathcal{B}_2^a$: $D(a, 3+a, 4+a, 2)$, $D(a, 2+a, 1+a, -2)$, $B(\infty_1, 4+a, 1+a, -1)$, $B(\infty_1, 3+a, 2+a, -2)$, $B(\infty_2, 2+a, 4+a, 2)$, $B(\infty_2, 1+a, 3+a, -2)$, $T(\infty_1, \infty_2, a, 2)$.

(4) $\mathcal{B}_k^a$ $(3 \le k \le \frac{tv-1}{2})$: $D(a, 1+a, 4+a, k)$, $D(a, 2+a, 3+a, k)$, $B(\infty_1, 1+a, 3+a, k)$, $B(\infty_1, 4+a, 2+a, k)$, $B(\infty_2, 3+a, 4+a, -k)$, $B(\infty_2, 2+a, 1+a, -k)$, $T(\infty_1, \infty_2, a, k)$.

(5) $\mathcal{B}_k^a$ $(\frac{tv+1}{2} \le k \le tv-3)$: $D(a, 1+a, 4+a, k)$, $D(a, 2+a, 3+a, k)$, $B(\infty_1, 3+a, 4+a, -k)$, $B(\infty_1, 2+a, 1+a, -k)$, $B(\infty_2, 1+a, 3+a, k)$, $B(\infty_2, 4+a, 2+a, k)$, $T(\infty_1, \infty_2, a, k)$.

(6) $\mathcal{B}_{tv-2}^a$: $D(a, 1+a, 3+a, 2)$, $D(a, 2+a, 4+a, -2)$, $B(\infty_1, 3+a, 4+a, 1)$, $B(\infty_1, 1+a, 2+a, -1)$, $B(\infty_2, 4+a, 1+a, 0)$, $B(\infty_2, 2+a, 3+a, -2)$, $T(\infty_1, \infty_2, a, -2)$.

(7) $\mathcal{B}_{tv-1}^a$: $D(a, 3+a, 4+a, 1)$, $D(a, 1+a, 2+a, -1)$, $B(\infty_1, 1+a, 4+a, 0)$, $B(\infty_1, 2+a, 3+a, 0)$, $B(\infty_2, 4+a, 2+a, 1)$, $B(\infty_2, 3+a, 1+a, -1)$, $T(\infty_1, \infty_2, a, 0)$.

Proof. It is easy to see that each $(Y \times Z_t, \mathcal{G}, \mathcal{B}_k^a)$, $k \in Z_{tv}, a \in Z_5$, is a 3-$GDD(t^{5v+2})$. On the other hand, denote $\mathcal{A} = \bigcup_{a \in Z_5} \bigcup_{k \in Z_{tv}} \mathcal{B}_k^a$, then it is not hard to verify that

(1) For $e \in Z_5$ and $k \in Z_{tv}$, $D(e, 1+e, 4+e, k)$, $D(e, 2+e, 3+e, k)$ and $T(\infty_1, \infty_2, e, k)$ each appears exactly once in $\mathcal{A}$.

(2) For $b \neq h \in Z_5$, and any $k \in Z_{tv}$, either $B(\infty_1, b, h, k)$ or $B(\infty_2, b, h, k)$ appears once in $\mathcal{A}$.

(3) For $b \neq h \in Z_5$, and $k \in Z_{tv}$, by Remark, $S(u, b, h, k)$ ($u \in \{\infty_1, \infty_2\}$) appears once in $\mathcal{A}$.

Thus, all $\mathcal{B}_k^a$ ($a \in Z_5$, $k \in Z_{tv}$) are mutually disjoint. Thereby, we obtain a large set $LS(t^{5v+2})$, indeed. □

Theorem 7.2.5. If an $LS(t^{v+2})$ exists, then an $LS(t^{5v+2})$ exists.

Proof. If t is even, Theorem 7.2.2 (2) cover this result. If t is odd, then v must be odd, thus the conclusion is true by Lemma 7.2.4. □

Furthermore, Chen and Lei have extended this construction and given following result.

Theorem 7.2.6. For prime $p \equiv 5 \pmod 8$, $p \leq 157$ and $u \geq 5$. If an $LS(t^{u+2})$ exists, then an $LS(t^{up+2})$ exists.

Through Theorems 7.2.3, 7.2.5 and 7.2.6, the sufficency for (LG1) and some subcases of (LG2) and (LG3) can be obtained. Recently, in [Teir 8], Luc Terilinck has affirmed whole sufficiency for cases (LG2) and (LG3). But for case (LG4), up to now we have none example. Using Teirlinck's method, we only need to show that there exists an $LS(6^{u+2})$ for any $u \in \{3 \times 2^n : n \geq 1\}$, then (LG4) would be sufficient. However, up to now, this problem is still open (even though Lei give $LS(6^{12k+8})$ for any nonnegative k, recently).

7.3 Spectrum of self-converse *MTS*

For a Mendelsolm triple system with index $\lambda = 1$, $MTS(v) = (X, \mathcal{B})$, define

$$\mathcal{B}^{-1} = \{B^{-1} : B \in \mathcal{B}\},$$

where $B^{-1} = \langle z, y, x\rangle$ if $B = \langle x, y, z\rangle$, and B^{-1} is called the *converse* of B. Obviously, $(X, \mathcal{B}^{-1})$ is also an $MTS(v)$, which is called the *converse* of $(X, \mathcal{B})$. If there exists an isormorphic mapping f from $(X, \mathcal{B})$ to $(X, \mathcal{B}^{-1})$, then the $MTS(v) = (X, \mathcal{B})$ is called *self-converse* and denoted by $SCMTS\ (v) = (X, \mathcal{B}, f)$. To prove a system $(X, \mathcal{B}, f)$ is self-converse we only need to show that $f(B)^{-1} \in \mathcal{B}$ for any $B \in \mathcal{B}$.

For what orders do self-converse MTS exist? The problem was posed by C. J. Colbourn and A. Rosa in their survey (as open problem 5, see [ColRo]). In this section, we will complete the existence spectrum of $SCMTS(v)$ (see [ChangYangKang]).

For convenience, we define some terminologies. Let $(X, \mathcal{B}, f)$ be an $SCMTS(v)$. For $x \in X$, if $f^k(x) = x$ but $f^s(x) \neq x$ when $s < k$, then denote $p_f(x) = k$. When $p_f(x) = 1$, we call x *fixed point* of the $SCMTS(v)$. If there exists an $SCMTS(u) = (Y, \mathcal{A}, g)$ such that $Y \subset X$, $\mathcal{A} \subset \mathcal{B}$ and $g = f|_Y$, then the sub-system is called a sub-$SCMTS(u)$ of the $SCMTS(v)$. An $SCMTS(v)$ is called

A-type if $p_f(x) \leq 2$ for any $x \in X$ and $|\{x \in X; p_f(x) = 1\}| \geq 2$;

B-type if $p_f(x) \leq 2$ for any $x \in X$ and there are a sub-$SCMTS(4)$ and another fixed point, which is not contained in the sub-$SCMTS(4)$;

C-type if $p_f(x) \leq 2$ for any $x \in X$ and there is a sub-$SCMTS(3)$, in which three points are all fixed.

Note: $MTS(3)$ and $MTS(4)$ are all unique (up to isomorphic), i.e. $MTS(3) = \{\langle a, b, c\rangle, \langle c, b, a\rangle\}$ and $MTS(4) = \{\langle u, v, s\rangle, \langle u, s, t\rangle, \langle u, t, v\rangle, \langle v, t, s\rangle\}$. But as self-converse system, $MTS(3)$ has six mappings, i.e. all permutations on $\{a, b, c\}$; and $MTS(4)$ has twelve mappings, i.e. all permutations in the form $(*)(*)(*,*)$ or $(*,*,*,*)$ on $\{u, v, s, t\}$. Here, the mapping f in C-type sub-$SCMTS(3)$ is stipulated for identical mapping, and the mapping g in B-type sub-$SCMTS(4)$ can be only the form $(*)(*)(*,*)$ by $p_f(x) \leq 2$ for any x. Obviously, by definition, C-type implies A-type.

7.3.1 Case $v \equiv 1,3$ and 4 (mod 6)

Theorem 7.3.1. For $v \equiv 1,3 \pmod 6$ there exists an $SCMTS(v)$, which is C-type.

Proof. It is well known that there exists a Steiner triple system $STS(v) = (X, \mathcal{A})$ for $v \equiv 1,3 \pmod 6$. Let $\mathcal{B} = \{\langle x,y,z\rangle, \langle z,y,x\rangle : \{x,y,z\} \in \mathcal{A}\}$ and f be an identical transformation on X. Then $(X, \mathcal{B}, f)$ is an $SCMTS(v)$. Obviously, this system is C-type. □

Lemma 7.3.2. If there exist both $SCMTS(m)$ and $SCMTS(n)$, then there exists an $SCMTS(mn)$. Furthermore, when $SCMTS(m)$ is A-type, if $SCMTS(n)$ is A-type (or C-type), then the obtained $SCMTS$ (mn) is A-type (or C-type) too.

Construction. Let X be a m-set and Y be a n-set. Given $SCMTS(m) = (X, \mathcal{A}, f)$ and $SCMTS(n) = (Y, \mathcal{B}, g)$. Construct a cyclic triple system J on the set $X \times Y$ as follows:

part 1. $\langle (x,u),(y,v),(z,w)\rangle$ with $\langle x,y,z\rangle \in \mathcal{A}$ and $\langle u,v,w\rangle \in \mathcal{B}$. This gives $mn(m-1)(n-1)/3$ cyclic triples.

part 2. $\langle (x,u),(x,v),(x,w)\rangle$ with $\langle u,v,w\rangle \in \mathcal{B}$ and $x \in X$. This gives $mn(n-1)/3$ cyclic triples.

part 3. $\langle (x,u),(y,u),(z,u)\rangle$ with $\langle x,y,z\rangle \in \mathcal{A}$ and $u \in Y$. This gives $mn(m-1)/3$ cyclic triples.

Define mapping $F : X \times Y \longrightarrow X \times Y$ such that $F(x,y) = (f(x), g(y))$ for $x \in X$ and $y \in Y$. Then $(X \times Y, J, F)$ is an $SCMTS(mn)$ as expected.

Proof. The system J contains

$$mn(m-1)(n-1)/3 + mn(n-1)/3 + mn(m-1)/3 = mn(mn-1)/3$$

cyclic triples, just the number as expected. For any $x \neq y \in X$, $u \neq v \in Y$, each ordered pair $((x,u),(x,v))$ appears in one block of part 2, each ordered pair $((x,u),(y,u))$ appears in one block of part 3 and each ordered pair $((x,u),(y,v))$ appears in one block of part 1. Thus $(X \times Y, J)$ is an $MTS(mn)$.

The triple $B = \langle (x,u),(y,v),(z,w)\rangle \in J$ implies both $\langle x,y,z\rangle \in \mathcal{A}$ and $\langle u,v,w\rangle \in \mathcal{B}$. Thus $\langle f(z), f(y), f(x)\rangle \in \mathcal{A}$ and $\langle g(w), g(v), g(u)\rangle \in \mathcal{B}$, thereby $\langle(f(z),g(w)),(f(y),g(v))\ ,(f(x),g(u))\ \rangle=\langle F(z,w),F(y,v),F(x,u)\rangle$ appears in part 1, i.e. $F(B)^{-1}\in J$. The triple $B=\langle(x,u),(x,v),(x,w)\rangle\in J$ implies $x\in X$ and $\langle u,v,w\rangle\in\mathcal{B}$. Thus $\langle g(w),\,g(v),g(u)\rangle\in\mathcal{B}$, thereby $\langle(f(x),g(w)),\ (f(x),g(v)),\ (f(x),g(u))\rangle=\langle F(x,w),F(x,v),F(x,u)\rangle$ appears in part 2, i.e. $F(B)^{-1}\in J$. Similar for the triple $\langle(x,u),(y,u),(z,u)\rangle$. Therefore $(X\times Y,J,F)$ is an $SCMTS(mn)$.

If $p_f(x)\le 2$ and $p_g(u)\le 2$, i.e. $f^2(x)=x$ and $g^2(u)=u$, for any $x\in X$ and $u\in Y$, then $F^2(x,u)=(f^2(x),g^2(u))=(x,u)$, i.e. $p_F(x,u)\le 2$, for any $(x,u)\in X\times Y$. Suppose $(X,\mathcal{A},f)$ is A-type, which has a fixed point x_0. If $(Y,\mathcal{B},g)$ is also A-type, in which $g(u_0)=u_0$ and $g(v_0)=v_0$, then (x_0,u_0) and (x_0,v_0) are fixed points of $(X\times Y,J,F)$. If $(Y,\mathcal{B},g)$ is C-type, in which there exists a sub-$SCMTS(3)=\{\langle a,b,c\rangle,\ \langle c,b,a\rangle\}$ and $g(a)=a$, $g(b)=b$, $g(c)=c$, then both $\langle(x_0,a),(x_0,b),(x_0,c)\rangle$ and $\langle(x_0,c),(x_0,b),(x_0,a)\rangle$ appear in part 2 of $(X\times Y,J,F)$ and $F(x_0,a)=(x_0,a)$, $F(x_0,b)=(x_0,b)$, $F(x_0,c)=(x_0,c)$. □

Theorem 7.3.3. For $v\equiv 4 \pmod 6$ there exists an $SCMTS(v)$, which is A-type, and is still B-type when $v>4$.

Construction. Let $v=6t+4=3(2t+1)+1$, $Z_{2t+1}=\{0,1,...,2t\}$, $Z_3=\{0,1,2\}$ and $\infty\notin Z_{2t+1}\times Z_3$. Construct a cyclic triple system $\mathcal{B}$ on the set $X=\{\infty\}\cup(Z_{2t+1}\times Z_3)$ as follows (The element (x,i) in $Z_{2t+1}\times Z_3$ is denoted by x_i, briefly):

part 1. $\bigcup\{MTS(4)$ on set $\{\infty\}\cup(\{x\}\times Z_3)$: $x\in Z_{2t+1}\}$;

part 2. $\langle(x+y)_i,(x-y)_i,x_{i+1}\rangle$ with $x\in Z_{2t+1}$, $1\le y\le t$ and $i\in Z_3$;

part 3. $\langle(x-y)_i,(x+y)_i,x_{i-1}\rangle$ with $x\in Z_{2t+1}$, $1\le y\le t$ and $i\in Z_3$.

Define mapping f on X as follows:

$$f(\infty)=\infty,f(x_0)=x_0,f(x_1)=x_2 \text{ and } f(x_2)=x_1 \text{ for any } x\in Z_{2t+1}.$$

Then $(X,\mathcal{B},f)$ is an $SCMTS(6t+4)$ as expected.

Proof. The system $\mathcal{B}$ contains

$$4(2t+1)+3t(2t+1)+3t(2t+1)=(6t+4)(6t+3)/3$$

cyclic triples, just the number as expected. Obviously, all ordered pairs (∞,x_i), (x_i,∞) and (x_i,x_j) are contained in triples of part 1, where

$x \in Z_{2t+1}$ and $i \neq j \in Z_3$. Now, Let us consider any ordered pair $P = (u_i, v_j)$, $u \neq v \in Z_{2t+1}$, $i, j \in Z_3$.

i) When $P = (u_i, v_i)$, there is a unique $x \in Z_{2t+1}$ such that $u+v = 2x$. If $u - x \leq t$, let $u - x = y$, then $u = x + y$, $v = x - y$ and P appears in part 2. If $u - x > t$, then $x - u = y \leq t$, thus $u = x - y$, $v = x + y$ and P appears in part 3.

ii) When $P = (u_i, v_{i+1})$, if $u - v = y \leq t$ then $u = x, v = x - y$ and P appears in part 3; if $u - v > t$, let $v - u = y$, then $u = x - y, v = x$ and P appears in part 2.

iii) When $P = (u_i, v_{i-1})$, if $u - v = y \leq t$ then $u = x + y$, $v = x$ and P appears in part 3; if $u - v > t$, let $v - u = y$, then $u = x, v = x + y$ and P appears in part 2.

Thus the system $(X, \mathcal{B})$ is an $MTS(6t + 4)$.

Under the mapping f, ∞ and all x_0 are fixed points, and $p_f(x_1) = 2$, $p_f(x_2) = 2$ for any $x \in Z_{2t+1}$. Obviously, each $MTS(4)$ in part 1 is a sub-$SCMTS(4)$. And the mapping f gives the following correspondences:

$$\begin{aligned}
\langle (x+y)_0, (x-y)_0, x_1 \rangle &\longleftrightarrow \text{the converse of } \langle (x-y)_0, (x+y)_0, x_2 \rangle, \\
\langle (x+y)_1, (x-y)_1, x_2 \rangle &\longleftrightarrow \text{the converse of } \langle (x-y)_2, (x+y)_2, x_1 \rangle, \\
\langle (x+y)_2, (x-y)_2, x_0 \rangle &\longleftrightarrow \text{the converse of } \langle (x-y)_1, (x+y)_1, x_0 \rangle.
\end{aligned}$$

Thus, the system $(X, \mathcal{B}, f)$ is an A-type $SCMTS(6t + 4)$. Furthermore, when $t \geq 1$, besides a sub-$SCMTS(4)$ in part 1 there is another fixed point, thereby the system is still B-type. □

Corollary 7.3.4. If there exists an $SCMTS(v)$, then there exist $SCMTS(3v)$, $SCMTS(4v)$ and $SCMTS(12v)$. And if the $SCMTS(v)$ is A-type (or C-type) then the $SCMTS(kv)$ above is the same type too, where $k = 3, 4, 12$.

Proof . There exists a C-type $SCMTS(3)$ which is also A-type, by Theorem 7.3.1. There exists an A-type $SCMTS(4)$ by Theorem 7.3.3. Thus there exists an A-type $SCMTS(3 \times 4)$ by Lemma 7.3.2. Thereby, the conclusion holds by Lemma 7.3.2. □

7.3.2 Case $v \equiv 0 \pmod{18}$

Lemma 7.3.5. There exists an A-type $SCMTS(18)$.

Construction. Let $X = \{a, b, c, d\} \cup Z_{14}$. Define mapping $X \longrightarrow X$ as follows: $f(a) = a$, $f(b) = b$, $f(c) = d$, $f(d) = c$ and $f(i) = 13 - i$ for $i \in Z_{14}$. The cyclic triple system $\mathcal{B}$ consists of the following triples:

part 1. $MTS(4)$ on $\{a, b, c, d\}$.

part 2. $\langle a, i, 13 - i\rangle$ and $\langle b, i, i + 8\rangle$, $i \in Z_{14}$.

part 3. $\langle c, i, i + 4\rangle$ and $\langle d, 9 - i, 13 - i\rangle$, $0 \le i \le 4$ and $10 \le i \le 11$.

part 4. $\langle c, i, i + 11\rangle$ and $\langle d, 2 - i, 13 - i\rangle$, $5 \le i \le 6$.

part 5. $\langle c, i, i + 3\rangle$ and $\langle d, 10 - i, 13 - i\rangle$, $7 \le i \le 9$.

part 6. $\langle c, 12, 13\rangle$, $\langle c, 13, 9\rangle$,
$\langle d, 0, 1\rangle$, $\langle d, 4, 0\rangle$.

part 7. $\langle i, i + 3, i + 8\rangle$, $\langle i, i + 10, i + 1\rangle$,
$\langle 13 - i, 5 - i, 10 - i\rangle$, $\langle 13 - i, 12 - i, 3 - i\rangle$, $i = 0, 11$.

part 8. $\langle i, i + 7, i + 2\rangle$ and $\langle 13 - i, 11 - i, 6 - i\rangle$, $i = 0, 5, 7, 12$.

part 9. $\langle j, j+i, j+2i\rangle$ and $\langle 13-2i-j, 13-i-j, 13-j\rangle$, $i = 1, j = 1, 3$; $i = 2, j = 3$; $i = 6, j = 0, 3$; $i = 9, j = 1$; $i = 11, j = 2$; and $i = 13, j = 4, 6$.

part 10. $\langle 5, 6, 9\rangle$, $\langle 0, 2, 9\rangle$, $\langle 1, 11, 4\rangle$, $\langle 2, 7, 4\rangle$,
$\langle 8, 4, 7\rangle$, $\langle 13, 4, 11\rangle$, $\langle 12, 9, 2\rangle$, $\langle 11, 9, 6\rangle$.

Proof . Firstly, $(X, \mathcal{B})$ is an $MTS(18)$ affirmatively, but this verification is tedious, which is deleted. For any triple B, it is easy to see that $f(B)^{-1}$ appears in the same part. Obviously, $f^2(x) = x$ holds for any $x \in X$ and a, b are fixed points. □

Theorem 7.3.6. For $v \equiv 0 \pmod{18}$ there exists an $SCMTS(v)$, which is A-type.

Proof . Let $v = 18t$, $t \ge 1$. Use induction method on t. When $t = 1$, there exists A-type $SCMTS(18)$ by Lemma 7.3.5. Suppose there exists an A-type $SCMTS(18r)$ for any $r < t$. Let us consider the next order $v = 18t$. When $t = 3r + 1$, $v = 18(3r + 1)$. Since $3r + 1 \equiv 1$ or $4 \pmod 6$, there exists an A-type $SCMTS(3r + 1)$ by Theorem 7.3.1 or 7.3.3. Since there exists an A-type $SCMTS(18)$ by Lemma 7.3.5, thus there exists an A-type $SCMTS(18(3r+1))$ by Lemma 7.3.2. When $t = 3r+2, v = 18(3r+2) = 9(6r+4)$. There exist both A-type $SCMTS(9)$ and $SCMTS(6r+4)$ by Thorem 7.3.1 and 7.3.3, thus there exists an A-type $SCMTS(9(6r+4))$ Lemma 7.3.2. Finally, when $t = 3r, v = 18 \cdot 3r = 3 \cdot 18r$. By induction hypothesis, since $r < t$, there exists an A-type $SCMTS(18r)$, thereby there exists an A-type $SCMTS(3 \cdot 18r)$ by Corollary 7.3.4. □

7.3.3 Stronger $SCMTS(6k+4)$

Lemma 7.3.7. If there exists an A-type $SCMTS(n+1)$ then there exists a C-type $SCMTS(3n+1)$.

Construction. Let $X = \{\infty\} \cup (Z_n \times Z_3)$, $\infty \notin Z_n \times Z_3$. By known condition, there exists an A-type $SCMTS(n+1) = (\{\infty\} \cup Z_n, \mathcal{A}, f)$, which contains two fixed points ∞ and 0, and $p_f(x) \leq 2$ for any $x \in Z_n \setminus \{0\}$. Now, construct a cyclic triple system $\mathcal{B}$ on the set X as follows:

part 1. $\langle (x,i),(y,i),(z,i) \rangle$ with $\langle x,y,z \rangle \in \mathcal{A}$ and $i \in Z_3$. (Whenever ∞ appears for x, y, z, omit the second coordinate i.)

part 2. $\langle (x,0),(y,1),(z,2) \rangle$ with $x, y, z \in Z_n$ and $x+y+z=0$.

part 3. $\langle (f(z),2),(f(y),1),(f(x),0) \rangle$ with $x, y, z \in Z_n$ and $x+y+z=0$.

Define $F(\infty) = \infty, F(x,i) = (f(x),i)$ for any $(x,i) \in Z_n \times Z_3$. Then $(X, \mathcal{B}, F)$ is an $SCMTS(3n+1)$ as expected.

Proof . $\mathcal{B}$ contains $3 \cdot \frac{n(n+1)}{3} + n^2 + n^2 = 3n(3n+1)/3$ triples, just the number as expected. All ordered pairs $(\infty,(x,i)), ((x,i),\infty)$ and $((x,i),(y,i))$ appears in cyclic triples of part 1. For ordered pair $P = ((x,i),(y,i+1))$, there is $z \in Z_n$ such that $x+y+z=0$, then P appears in triple $\langle (x,i),(y,i+1),(z,i-1) \rangle$ of part 2. For $P = ((x,i),(y,i-1))$, let $u = f(x), v = f(y), w = -(u+v)$ and $z = f(w)$. Since $f(u) = f^2(x) = x$ and $f(v) = f^2(y) = y$, thus P appears in triple $\langle (x,i),(y,i-1),(z,i+1) \rangle$ of part 3.

Triple $A = \langle (x,i), (y,i),(z,i) \rangle \in \mathcal{B}$ implies $\overline{A} = \langle x,y,z \rangle \in \mathcal{A}$, so $f(\overline{A})^{-1} = \langle f(z), f(y), f(x) \rangle \in \mathcal{A}$, thereby $\langle (f(z),i), (f(y),i), (f(x),i)$

$\rangle \in$ part 1, i.e. $F(A)^{-1} \in \mathcal{B}$. Triple $B = \langle(x,0),\ (y,1),\ (z,2)\rangle \in \mathcal{B}$ implies $x+y+z=0$, so $\langle(f(z),2), (f(y),1), (f(x),0)\ \rangle \in$ part 3, i.e. $F(B)^{-1} \in \mathcal{B}$. For the triple $C = \langle(x,2), (y,1),\ (z,0)\rangle \in \mathcal{B}$, let $u = f(x), v = f(y)$ and $w = f(z)$, then $f(u) = f^2(x) = x$, $f(v) = f^2(y) = y$ and $f(w) = f^2(z) = z$. Thus, $C = \langle(x,2), (y,1),\ (z,0)\rangle = \langle\ (f(u),2), (f(v),1), (f(w),0)\rangle \in B$ implies $u+v+w=0$ and $\langle(w,0), (v,1), (u,2)\rangle = \langle(f(z),0), (f(y),1), (f(x),2)\ \rangle \in$ part 2, that is $f(C)^{-1} \in \mathcal{B}$.

Finally, since $f(0) = 0$ and $0+0+0=0$, thus $F(0,i) = (0,i), i \in Z_3$ and the $(X, \mathcal{B}, f)$ contains a sub-$SCMTS(3) = \{\langle(0,0),(0,1),(0,2)\rangle, \langle(0,2),(0,1),(0,0)\rangle\}$ by part 2 and part 3 . And for any element $(x,i) \in Z_n \times Z_3$, $F^2(x,i) = (f^2(x),i) = (x,i)$, thus $p_F((x,i)) \leq 2$. The element ∞ is another fixed point. Thus, $(X, \mathcal{B}, F)$ is C-type. □

Lemma 7.3.8. If there exists a B-type $SCMTS(n+4)$ then there exists a C-type $SCMTS(3n+4)$, for $n > 0$.

Construction. Given a B-type $SCMTS(n+4) = (\{u,v,s,t\} \cup Z_n, \mathcal{A}, f)$, which contains a sub-$SCMTS(4)$ on the sub-set $\{u,v,s,t\}$, and satisfies $p_f(x) \leq 2$ for any $x \in Z_n$, where $x = 0$ is fixed point. Construct a cyclic triple system $\mathcal{B}$ on set $X = \{u,v,s,t\} \cup (Z_n \times Z_3)$ as follows:

part 1. sub-$SCMTS(4)$ on $\{u,v,s,t\}$, say D.

part 2. $\langle(x,i),(y,i),(z,i)\rangle$ with $\langle x,y,z\rangle \in \mathcal{A} \backslash D$ and $i \in Z_3$. (whenever u,v,s,t appears for x,y,z, omit the second coordinate i.)

part 3. $\langle(x,0),(y,1),(z,2)\rangle$ and $\langle(f(z),2),(f(y),1),(f(x),0)\rangle$ with $x,y,\ z \in Z_n$ and $x+y+z=0$.

Define $F(u) = f(u), F(v) = f(v), F(s) = f(s), F(t) = f(t)$ and $F(x,i) = (f(x),i)$ for any $(x,i) \in Z_n \times Z_3$. Then $(X, \mathcal{B}, F)$ is an $SCMTS(3n+4)$ as expected.

Proof. It is not difficult to show that $(X, \mathcal{B})$ is an $MTS(3n+4)$, similar to Lemma 7.3.7. If B is a triple from $\{u,v,s,t\}$, then $F(B)^{-1} = f(B)^{-1}$ appears in part 1. As sub-$SCMTS(4)$ of $\mathcal{A}$, D is also a sub-$SCMTS(4)$ of $\mathcal{B}$, thereby $p_F(z) \leq 2$ for $z = u,v,s,t$. Other statements of the proof are almost the same to that in Lemma 7.3.7. □

Lemma 7.3.9. If there exists a C-type $SCMTS(n+3)$ then there exists a C-type $SCMTS(3n+3)$.

Construction. Given a C-type $SCMTS(n+3) = (\{a,b,c\} \cup Z_n, \mathcal{A}, f)$, which contains a sub-$SCMTS(3)$, $D = \{\langle a,b,c\rangle, \langle c,b,a\rangle\}$, and satisfies $p_f(x) \le 2$ for any $x \in Z_n$. Let us construct a cyclic triple system $\mathcal{B}$ on set $X = \{a,b,c\} \cup (Z_n \times Z_3)$ as follows:

part 1. D.

part 2. $\langle (x,i),(y,i),(z,i)\rangle$ with $\langle x,y,z\rangle \in \mathcal{A} \setminus D$ and $i \in Z_3$. (Whenever a,b,c appears for x,y,z, omit the second coordinate i.)

part 3. $\langle (x,0),(y,1),(z,2)\rangle$ and $\langle (f(z),2),(f(y),1),(f(x),0)\rangle$ with $x,y,\ z \in Z_n$ and $x+y+z=0$.

Define $F(a) = a, F(b) = b, F(c) = c$ and $F(x,i) = (f(x),i)$ for any $(x,i) \in Z_n \times Z_3$. Then $(X,\mathcal{B},F)$ is an $SCMTS(3n+3)$ as expected.

Proof. Similar to Lemma 7.3.7 and Lemma 7.3.8. □

Lemma 7.3.10. There exists a C-type $SCMTS(v)$ for $v \equiv 4$ $(mod$ $6)$ and $v > 4$.

Proof. Let $v = 6k+4$, all posibilities are classified as follows.

Case 1. $k = 3t$, $t \ge 1$, then $v = 18t+4 = 3 \times 6t + 4$.

There exists a B-type $SCMTS$ $(6t+4)$ by Theorem 7.3.3, thus there exists a C-type $SCMTS(3 \times 6t+4))$ by Lemma 7.3.8.

Case 2. $k = 3t+1$, then $v = 18t+10 = 3 \times (6t+3)+1$. There exists an A-type $SCMTS(6t+3+1)$ by Theorem 7.3.3, thus there exists a C-type $SCMTS(3 \times (6t+3)+1)$ by Lemma 7.3.7.

Case 3. $k = 3t+2$, then $v = 18t+16$.

Subcase 1. $t = 2s$, then $v = 36s+16 = 4(9s+4)$. There exists a C-type $SCMTS(9s+4)$ by Theorem 7.3.1 (when s odd) or Case 1 (when s even), thus there exists a C-type $SCMTS(4(9s+4))$ by Corollary 7.3.4.

Subcase 2. $t = 2s+1$, then $v = 36s+34 = 3(12s+11)+1$. If there exists an A-type $SCMTS(12s+12)$ then there exists a C-type $SCMTS(v)$ by Lemma 7.3.7. Below, we will discuss the existence of A-type $SCMTS(12s+12)$. Firstly, since there exist C-type $SCMTS(18p+4)$ and $SCMTS(18p+10)$ by Case 1 and Case 2; by Lemma 7.3.9, there exist C-type.

$$SCMTS(54p+6) \text{ and } SCMTS(54p+24). \qquad (*)$$

Now, we can get the existence of A-type $SCMTS(12s+12)$ (classified by $s = 6m + i,\ i \in Z_6$):

(i) $72m + 12 = 12(6m + 1)$: by Theorem 7.3.1 and Corollary 7.3.4.

(ii) $72m + 24$, let $m = 3p + i$ $(i \in Z_3)$:

$$\begin{cases} 216p + 24 = 4(54p + 6) & \text{by } (*) \text{ and Corollary 7.3.4;} \\ 216p + 96 = 4(54p + 24) & \text{by } (*) \text{ and Corollary 7.3.4;} \\ 216p + 168 = 54(4p + 3) + 6 & \text{by } (*). \end{cases}$$

(iii) $72m + 36 = 4(18m + 9)$: by Theorem 7.3.1 and Corollary 7.3.4.

(iv) $72m + 48 = 12(6m + 4)$: by Theorem 7.3.3 and Corollary 7.3.4.

(v) $72m + 60 = 4(18m + 15)$: by Theorem 7.3.1 and Corollary 7.3.4.

(vi) $72m + 72 = 4(18m + 18)$: by Theorem 7.3.6 and Corollary 7.3.4.

This completes the proof. □

7.3.4 The conclusion

Theorem 7.3.11. There exists an $SCMTS(v)$ if and only if $v \equiv 0$ or $1 \pmod 3$ and $v \neq 6$.

Proof . The necessity is obvious due to the spectrum of $MTS(v)$. When $v \equiv 1$ or $3 \pmod 6$ and $v \equiv 4 \pmod 6$, the sufficiency holds by Theorem 7.3.1 and Theorem 7.3.3. When $v \equiv 0 \pmod 6$, if $v = 18t$ the conclusion has already been given by Theorem 7.3.6; if $v = 18t+6$, $t \geq 1$, there exists a C-type $SCMTS(6t+4)$ by Lemma 7.3.10, thus there exists an $SCMTS(18t + 6)$ by Lemma 7.3.9; if $v = 18t + 12 = 3(6t + 4)$, then the conclusion can be gotten by Theorem 7.3.3 and Corollary 7.3.4. □

Recently, Zhang Jie obtains the existence spectrum of $SCMTS(v, \lambda)$ without repeated triples for any index λ.

7.4 Spectrum of self-converse DTS

For a transitive triple system with index $\lambda = 1$, $DTS(v) = (X, \mathcal{B})$, define

$$\mathcal{B}^{-1} = \{B^{-1} : B \in \mathcal{B}\},$$

where $B^{-1} = (z, y, x)$ is called the *converse* of $B = (x, y, z)$. Obviously, $(X, \mathcal{B}^{-1})$ is also a $DTS(v)$, which is called *converse* of $(X, \mathcal{B})$. If there exists an isormorphic mapping f from $(X, \mathcal{B})$ to $(X, \mathcal{B}^{-1})$, then the $DTS(v) = (X, \mathcal{B})$ is called *self-converse* and denoted by $SCDTS(v) = (X, \mathcal{B}, f)$. To prove a system $(X, \mathcal{B}, f)$ is self-converse we only need to show that $f(B)^{-1} \in \mathcal{B}$ for any $B \in \mathcal{B}$.

For what orders do self-converse DTS exist? The problem was posed by C. J. Colbourn and A. Rosa in their survey (as open problem 4, see [ColRo]). In this section, we will complete the existence spectrum of $SCDTS(v)$ (see [KangChangYang]).

For convenience, we define some terminologies. Let $(X, \mathcal{B}, f)$ be an $SCDTS(v)$. For $x \in X$, if $f^k(x) = x$ but $f^s(x) \neq x$ when $s < k$, then denote $p_f(x) = k$. When $p_f(x) = 1$, we call x *fixed point* of the $SCDTS(v)$. If there exists an $SCDTS(u) = (Y, \mathcal{A}, g)$ such that $Y \subset X$, $\mathcal{A} \subset \mathcal{B}$ and $g = f|_Y$, then the sub-system is called a sub-$SCDTS(u)$ of the $SCDTS(v)$. An $SCDTS(v)$ is said to be

A-type if $p_f(x) \leq 2$ for any $x \in X$ and $|\{x \in X; p_f(x) = 1\}| \geq 2$;

B-type if $p_f(x) \leq 2$ for any $x \in X$ and there are a sub-$SCDTS(4)$ and another fixed point, which is not contained in the sub-$SCDTS(4)$;

C-type if $p_f(x) \leq 2$ for any $x \in X$ and there is a sub-$SCDTS(3)$, in which three points are all fixed.

Note: $DTS(3) = \{(a, b, c),\ (c, b, a)\}$ is unique (up to isomorphic), as self-converse system it has two mappings: $f_1 = (a)(b)(c)$ and $f_2 = (b)(a, c)$. There are three nonisomorphic $DTS(4)$. Among them only one is self-converse, i.e. $\{(u, v, s),\ (w, s, v),\ (s, w, u),\ (v, u, w)\}$, which has four mappings: $g_1 = (u)(s)(v, w)$, $g_2 = (w)(v)(u, s)$, $g_3 = (u, v, s, w)$ and $g_4 = (w, s, v, u)$. Here, the mapping f in C-type sub-$SCDTS(3)$ is stipulated for f_1, and the mapping g in B-type sub-$SCDTS(4)$ can be only g_1 or g_2, by $p_f(x) \leq 2$ for any x. Obviously, C-type implies A-type by definition.

7.4.1 Case $v \equiv 1, 3$ and $4 \pmod 6$

Theorem 7.4.1. For $v \equiv 1, 3 \pmod 6$ there exists an $SCDTS(v)$, which is C-type.

Proof. It is well known that there exists Steiner triple system $STS(v) = (X, \mathcal{A})$ for $v \equiv 1, 3 \pmod 6$. Assign an order for each triple of $\mathcal{B}$. Let $\mathcal{B} = \{(x, y, z), (z, y, x); \{x, y, z\} \in \mathcal{A}\}$ and f be an identical transformation on X. Then $(X, \mathcal{B}, f)$ is an $SCDTS(v)$. Obviously, this system is C-type. □

Lemma 7.4.2. If there exists an $SCDTS(m)$ then there exists an $SCDTS(mn)$ for $n \equiv 1, 3 \pmod 6$, $n > 1$. Furthermore, if $SCDTS(m)$ is A-type, then the obtained $SCDTS(mn)$ is C-type.

Construction. Let X be an m-set and Y be an n-set. Given $SCDTS$ $(m) = (X, \mathcal{A}, f)$ and $STS(n) = (Y, \mathcal{B})$, where each unordered triple $\{u, v, w\} \in \mathcal{B}$ is assigned an order and denoted by $\{u, v, w\}_0$, for example $\{u, v, w\}_0 = (u, v, w)$. Construct a directed triple system J on the set $X \times Y$ as follows:

part 1. $((x, u), (y, v), (z, w))$ with $(x, y, z) \in \mathcal{A}$ and $\{u, v, w\} \in \mathcal{B}$. This gives $mn(m-1)(n-1)/3$ transitive triples.

part 2. $((x, u), (x, v), (x, w))$ and $((x, w), (x, v), (x, u))$ with $\{u, v, w\} \in \mathcal{B}$, $\{u, v, w\}_0 = (u, v, w)$ and $x \in X$. This gives $mn(n-1)/3$ transitive triples.

part 3. $((x, u), (y, u), (z, u))$ with $(x, y, z) \in \mathcal{A}$ and $u \in Y$. This gives $mn(m-1)/3$ transitive triples.

Define mapping $F: X \times Y \longrightarrow X \times Y$ such that $F(x, u) = (f(x), u)$ for $x \in X$ and $u \in Y$. Then $(X \times Y, J, F)$ is an $SCDTS(mn)$ as expected.

Proof. The system J contains

$$mn(m-1)(n-1)/3 + mn(n-1)/3 + mn(m-1)/3 = mn(mn-1)/3$$

transitive triples, just the number as expected. For any $x \neq y \in X$,

$u \neq v \in Y$, each ordered pair $((x,u),(x,v))$ appears in one block of part 2, each ordered pair $((x,u),(y,u))$ appears in one block of part 3 and each ordered pair $((x,u),(y,v))$ appears in one block of part 1. Thus $(X \times Y, \mathcal{J})$ is a $DTS(mn)$.

The triple $B = ((x,u),(y,v),(z,w)) \in \mathcal{J}$ implies both $(x,y,z) \in \mathcal{A}$ and $\{u,v,w\} \in \mathcal{B}$. Thus $(f(z), f(y), f(x)) \in \mathcal{A}$, thereby $((f(z),w), (f(y),v), (f(x), u)) = (F(z,w), F(y,v), F(x,u))$ appears in part 1, i.e. $F(B)^{-1} \in \mathcal{J}$. The triple $B = ((x,u),(x,v),(x,w)) \in \mathcal{J}$ implies $\{u,v,w\} \in \mathcal{B}$ and $\{u,v,w\}_0 = (u,v,w)$ or (w,v,u). Thus $((f(x),w),(f(x),v), (f(x),u)) = (F(x,w), F(x,v), F(x,u))$ appears in part 2, i.e. $F(B)^{-1} \in \mathcal{J}$. Similar for the triple $((x,u),(y,u),(z,u))$. Therefore $(X \times Y, \mathcal{J}, F)$ is an $SCDTS(mn)$.

If $p_f(x) \leq 2$, i.e. $f^2(x) = x$, for any $x \in X$ and $u \in Y$, then $F^2(x,u) = (f^2(x),u) = (x,u)$, i.e. $p_F(x,u) \leq 2$, for any $(x,u) \in X \times Y$. Suppose $(X, \mathcal{A}, f)$ is A-type, which has a fixed point x_0. Then both $((x_0,u),(x_0,v),(x_0,w))$ and $((x_0,w),(x_0,v),(x_0,u))$ appear in part 2 of $(X \times Y, \mathcal{J}, F)$ and $(x_0,u),(x_0,v)$ and (x_0,w) are all fixed. □

Theorem 7.4.3. For $v \equiv 4 \pmod 6$ there exists an $SCDTS(v)$, which is A-type, and is still B-type when $v > 4$.

Construction. Let $v = 6t+4 = 3(2t+1)+1$, $Z_{2t+1} = \{0,1,\ldots,2t\}$, $Z_3 = \{0,1,2\}$ and $\infty \notin Z_{2t+1} \times Z_3$. Construct a transitive triple system $\mathcal{B}$ on the set $X = \{\infty\} \cup (Z_{2t+1} \times Z_3)$ as follows (The element (x,i) in $Z_{2t+1} \times Z_3$ is denoted by x_i, briefly):

part 1. (∞, x_2, x_0), (x_2, ∞, x_1), (x_0, x_1, ∞), (x_1, x_0, x_2) with $x \in Z_{2t+1}$;

part 2. $((x+y)_i, x_{i+1}, (x-y)_i)$ with $x \in Z_{2t+1}$, $1 \leq y \leq t$ and $i \in Z_3$;

part 3. $((x-y)_i, x_{i-1}, (x+y)_i)$ with $x \in Z_{2t+1}$, $1 \leq y \leq t$ and $i \in Z_3$.

Define mapping f on X as follows:

$$f(\infty) = \infty, f(x_0) = x_0, f(x_1) = x_2 \text{ and } f(x_2) = x_1 \text{ for any } x \in Z_{2t+1}.$$

Then $(X, \mathcal{B}, f)$ is an $SCDTS(6t+4)$.

Proof . The system $\mathcal{B}$ contains

$$4(2t+1)+3t(2t+1)+3t(2t+1)=(6t+4)(6t+3)/3$$

transitive triples, just the number expected. Obviously, all ordered pairs (∞, x_i), (x_i, ∞) and (x_i, x_j) are contained in triples of part 1, where $x \in Z_{2t+1}$ and $i \neq j \in Z_3$. Now, Let us consider any ordered pair $P=(u_i, v_j)$, $u \neq v \in Z_{2t+1}$, $i, j \in Z_3$.

i) When $P=(u_i, v_i)$, there is a unique $x \in Z_{2t+1}$ such that $u+v=2x$. If $u-x \leq t$, let $u-x=y$, then $u=x+y$, $v=x-y$ and P appears in part 2. If $u-x>t$, then $x-u=y \leq t$, thus $u=x-y$, $v=x+y$ and P appears in part 3.

ii) When $P=(u_i, v_{i+1})$, if $u-v=y \leq t$ then $u=x+y, v=x$ and P appears in part 2; if $u-v>t$, let $v-u=y$, then $u=x, v=x+y$ and P appears in part 3.

iii) When $P=(u_i, v_{i-1})$, if $u-v=y \leq t$ then $u=x$, $v=x-y$ and P appears in part 2; if $u-v>t$, let $v-u=y$, then $u=x-y, v=x$ and P appears in part 3.

Thus the system $(X, \mathcal{B})$ is a $DTS(6t+4)$.

Under the mapping f, ∞ and all x_0 are fixed points, and $p_f(x_1)=2$, $p_f(x_2)=2$ for any $x \in Z_{2t+1}$. Obviously, each $DTS(4)$ in part 1 is a sub-$SCDTS(4)$. And the mapping f gives the following correspondences:

$$((x+y)_0, x_1, (x-y)_0) \longleftrightarrow \text{the converse of } ((x-y)_0, x_2, (x+y)_0),$$
$$((x+y)_1, x_2, (x-y)_1) \longleftrightarrow \text{the converse of } ((x-y)_2, x_1, (x+y)_2),$$
$$((x+y)_2, x_0, (x-y)_2) \longleftrightarrow \text{the converse of } ((x-y)_1, x_0, (x+y)_1).$$

Thus, the system $(X, \mathcal{B}, f)$ is an A-type $SCDTS(6t+4)$. Furthermore, when $t \geq 1$, besides a sub-$SCDTS(4)$ in part 1 there is another fixed point, thereby the system is still B-type. □

7.4.2 Case $v \equiv 0 \pmod{18}$

Lemma 7.4.4. There exists an A-type $SCDTS(18)$.

Construction. Let $X=\{a, b, c, d\} \cup Z_{14}$. Define mapping $f: X \longrightarrow X$ as follows: $f(a)=a$, $f(b)=b, f(c)=d, f(d)=c$ and $f(i)=13-i$ for $i \in Z_{14}$. The transitive triple system $\mathcal{B}$ consists of the following triple:

part 1. (a,d,b), (d,a,c), (b,c,a), (c,b,d).

part 2. $(i,a,13-i)$ and $(i,b,i+8)$, $i \in Z_{14}$.

part 3. $(i,c,i+4)$ and $(9-i,d,13-i)$, $0 \le i \le 4$ and $10 \le i \le 11$.

part 4. $(i,c,i+11)$ and $(2-i,d,13-i)$, $5 \le i \le 6$.

part 5. $(i,c,i+3)$ and $(10-i,d,13-i)$, $7 \le i \le 9$.

part 6. $(12,c,13),(13,c,9)$,
$(0\ ,d,1\ ),(4\ ,d,0)$.

part 7. $(i,i+7,i+2)$ and $(11-i,6-i,13-i)$, $i=0,5,7,12$.

part 8. $(j,j+i,j+2i)$ and $(13-j-2i,13-j-i,13-j)$, $i=1,j=1,3;i=5,j=0,1;i=6,j=0,2;i=12,j=7$; and $i=13,j=4,6$.

part 9. (0, 11, 3), (2, 9, 1), (3, 1, 0), (3, 12, 9), (6, 9, 2), (7, 8, 4), (9, 11, 0), (10, 5, 1), (10, 2, 13), (12, 4, 11), (13, 12, 10), (4, 1, 10), (11, 4, 7), (9, 5, 6), (13, 2, 4), (12, 8, 3).

Proof . Firstly, $(X,\mathcal{B})$ is a $DTS(18)$ affirmatively, but this verification is tedious, which is deleted. For any triple B, it is easy to see that $f(B)^{-1}$ appears in the same part. Obviously, $f^2(x)=x$ holds for any $x \in X$ and a,b are fixed points. □

Lemma 7.4.5. There exists an A-type $SCDTS(24)$.

Construction. Let Z_7 be a residue class ring module 7. Define a mapping f on Z_7 as follows: 0 is fixed and $1 \longleftrightarrow 4, 2 \longleftrightarrow 5, 3 \longleftrightarrow 6$. Let $X=\{a,b,c\}\cup(Z_7\times I_3)$, where $\{a,b,c\}\cap(Z_7\times I_3)=\phi$. Let $Q\{u,v,s,t\}=\{(u,t,v),\ (t,u,s),\ (v,s,u),\ (s,v,t)\}$ be a $DTS(4)$ on the set $\{u,v,s,t\}$ and $T\{u,v,s\}$ be a $DTS(3)$ on the set $\{u,v,s\}$, where $u,v,s,t \in X$. Construct a transitrve triple system $\mathcal{B}$ on the set X as follows:

part 1. $T\{a,b,c\}$;

part 2. $\bigcup\{T\{a,(2,i),(6,i)\},T\{a,(5,i),(3,i)\}:\ i \in I_3\}$;

part 3. $\bigcup\{T\{b,(1,i),(6,i)\},T\{b,(4,i),(3,i)\}:\ i \in I_3\}$;

part 4. $\bigcup\{T\{c,(1,i),(5,i)\},T\{c,(4,i),(2,i)\}:\ i \in I_3\}$;

part 5. $\bigcup\{T\{(1,i),(2,i),(3,i)\},T\{(4,i),(5,i),(6,i)\}:\ i \in I_3\}$;

part 6. $\bigcup\{Q\{a,(0,i),(1,i),(4,i)\},Q\{b,(0,i),(2,i),(5,i)\},Q\{c,(0,i),(3,i),(6,i)\}:\ i \in I_3\}$;

part 7. $((x,0),(y,1),(-x-y,2))$ and $((f(-x-y),2),(f(y),1),(f(x),0))$

with $x, y \in Z_7$.
Define mapping F on the set X as follows: a, b, c are fixed and $F(x, i) = (f(x), i)$ for $x \in Z_7$ and $i \in I_3$. Then $(X, \mathcal{B}, F)$ is an A-type $SCDTS(24)$.

Proof. It is easy to count that

$$|\mathcal{B}| = 2 + 12 \times 4 + 12 \times 3 + 7^2 \times 2 = 184 = \frac{24 \times 23}{3}.$$

Obviously, $f^2(x) = x$ holds for any $x \in Z_7$, thus $F^2(u) = u$ for any $u \in X$. Other details of the proof are omited. □

Lemma 7.4.6. If there exists an $SCDTS(v)$ then there exists an $SCDTS$ $(4v)$. Furthermore, when $SCDTS(v)$ is A-type (or C-type) then the obtained $SCDTS(4v)$ is also A-type (or C-type).

Construction. Given that $SCDTS(v) = (I_v, \mathcal{A}, f)$ and let

$$\mathcal{B}_c = \{\langle 0,1,2\rangle, \langle 1,0,3\rangle, \langle 2,3,0\rangle, \langle 3,2,1\rangle\},$$

$$\mathcal{B}_t = \{(0,2,1), (1,3,0), (2,0,3), (3,1,2)\}.$$

Define $g(0) = 0, g(1) = 1, g(2) = 3$, and $g(3) = 2$. Then $(I_4, \mathcal{B}_c, g)$ and $(I_4, \mathcal{B}_t, g)$ are $SCMTS(4)$ and $SCDTS(4)$, respectively. Now, construct a transitive triple system J on set $I_v \times I_4$ as follows:

part 1. $((x, i), (x, j), (x, k))$ with $x \in I_v$ and $(i, j, k) \in \mathcal{B}_t$.
part 2. $((x, i), (y, i), (z, i))$ with $(x, y, z) \in \mathcal{A}$ and $i \in I_4$.
part 3. $((x, i), (y, j), (z, k))$ with $(x, y, z) \in \mathcal{A}$ and $\langle i, j, k\rangle \in \mathcal{B}_c$.

Define $F(x, i) = (f(x), g(i))$, then $(I_v \times I_4, J, F)$ is a so $DTS(4v)$.

Proof. The system contains

$$4v + 4 \cdot \frac{v(v-1)}{3} + 12 \cdot \frac{v(v-1)}{3} = \frac{4v(4v-1)}{3}$$

transitive triples, just the number as expected. For any $x \neq y \in I_v$ and $i \neq j \in I_4$, the ordered pairs $((x, i), (x, j)), ((x, i), (y, i))$ and $((x, i), (y, j))$ appear in part 1, part 2 and part 3, respectively.

The triple $A = ((x, i), (x, j), (x, k)) \in J$ implies $(i, j, k) \in \mathcal{B}_t$. Thus $(g(k), g(j), g(i)) \in \mathcal{B}_t$ and $((f(x), g(k)), (f(x), g(j)), (f(x), g(i))) = (F(x, k), F(x, j), F(x, i))$ appears in part 1, i.e. $F(A)^{-1} \in J$. The triple $B =$

$((x,i), (y,i), (z,i)) \in \mathcal{J}$ implies $(x,y,z) \in \mathcal{A}$. Thus $(f(z), f(y), f(x)) \in \mathcal{A}$ and $((f(z), g(i)), (f(y), g(i)), (f(x), g(i))) = (F(z,i), F(y,i), F(x,i))$ appears in part 2, i.e. $F(B)^{-1} \in \mathcal{J}$. The triple $C = ((x,i),(y,j),(z,k)) \in \mathcal{J}$ implies $(x,y,z) \in \mathcal{A}$ and $\langle i,j,k\rangle \in \mathcal{B}_c$. Thus $(f(z),f(y),f(x)) \in \mathcal{A}$, $\langle g(k),g(j),g(i)\rangle \in \mathcal{B}_c$ and $((f(z),f(k)),(f(y),g(j)),(f(x),g(i)))=(F(z,k), F(y,j), F(x,i))$ appears in part 3, i.e. $F(C)^{-1} \in \mathcal{F}$.

Finally, since $g^2(i) = i$ for any $i \in I_4$, if $f^2(x) = x$ for any $x \in I_v$ then $F^2(x,i) = (f^2(x),g^2(i)) = (x,i)$. For A-type $SCDTS(v)$, let x be a fixed point, then both $(x,0)$ and $(x,1)$ are fixed points of the $SCDTS(4v)$ obtained. For C-type $SCDTS(v)$, suppose $f(a) = a, f(b) = b, f(c) = c$ and $(a,b,c),(c,b,a) \in \mathcal{A}$, then $((a,0),(b,0),(c,0))$, $((c,0),(b,0),(a,0)) \in \mathcal{J}$ and $(a,0),(b,0),(c,0)$ are fixed point of the $SCDTS(4v)$. □

Corollary 7.4.7. If there exists an A-type (or C-type) $SCDTS(v)$ then there exists an A-type (or C-type) $SCDTS(4^k v)$ and $SCDTS$ $(12v)$ for any integer $k > 0$.

Proof. Note that $12v = 4(3v)$, the conclusions follows by Theorem 7.4.1, Lemma 7.4.2 and Lemma 7.4.6. □

Theorem 7.4.8. For $v \equiv 0 \pmod{18}$ there exists an $SCDTS(v)$, which is A-type.

Proof. Let $v = 18n$ and classify by $n = 4k + i$, $i \in Z_4$.

Case 1. $v = 18(4k+1)$, $k = 3t+j$, $j \in Z_3$. If $v = 18(12t+1)$, or $18(12t+9)$ by Theorem 7.4.1, Lemma 7.4.4, and Lemma 7.4.2. If $v = 18(12t+5) = 9(24t+10)$, by Theorem 7.4.1, Theorem 7.4.3 and Lemma 7.4.2.

Case 2. $v = 18(4k+2) = 4(18k+9)$, by Theorem 7.4.1 and Lemma 7.4.6.

Case 3. $v = 18(4k+3)$, $k = 3t+j$, $j \in Z_3$. If $v = 18(12t+3)$ or $18(12t+7)$, by Theorem 7.4.1, Lemma 7.4.4 and Lemma 7.4.2. If $v = 18(12t+11) = 9(24t+22)$, by Theorem 7.4.1, Theorem 7.4.3 and Lemma 7.4.2.

Case 4. $v = 18(4k) = 2^{m+3}3^{n+2}u$, where $m \geq 0$, $n \geq 0$ and $u \equiv \pm 1 \pmod 6$. If $m = 2s$, then $v = 4^s \cdot 24(3^{n+1}u)$, by Theorem 7.4.1,

Lemma 7.4.5, Lemma 7.4.2 and Corollary 7.4.7. If $m = 2s+1$, then $v = 4^{s+2}(3^{n+2}u)$, by Theorem 7.4.1 and Corollary 7.4.7. Note that $3^n u \equiv 3 \pmod 6$ for any integer $n > 0$ and $u \equiv \pm 1 \pmod 6$. □

7.4.3 Stronger $SCDTS(6k+4)$

Lemma 7.4.9. If there exists an A-type $SCDTS(n+1)$ then there exists a C-type $SCDTS(3n+1)$.

Construction. Let $X = \{\infty\} \cup (Z_n \times Z_3)$, $\infty \notin Z_n \times Z_3$. Suppose that there exists an A-type $SCDTS(n+1) = (\{\infty\} \cup Z_n, \mathcal{A}, f)$, which contains two fixed points ∞ and 0, and $p_f(x) \le 2$ for any $x \in Z_n \setminus \{0\}$. Now, construct a transitive triple system $\mathcal{B}$ on the set X as follows:

part 1. $((x,i),(y,i),(z,i))$ with $(x,y,z) \in \mathcal{A}$ and $i \in Z_3$. (whenever ∞ appears for x, y, z, omit the second coordinate i.)

part 2. $((x,0),(y,1),(z,2))$ with $x, y, z \in Z_n$ and $x+y+z = 0$.

part 3. $((f(z),2),(f(y),1),(f(x),0))$ with $x, y, z \in Z_n$ and $x+y+z = 0$.

Define $F(\infty) = \infty, F(x,i) = (f(x),i)$ for any $(x,i) \in Z_n \times Z_3$. Then $(X, \mathcal{B}, F)$ is an $SCDTS(3n+1)$.

Proof. The system $\mathcal{B}$ contains $3 \cdot \frac{n(n+1)}{3} + n^2 + n^2 = 3n(3n+1)/3$ triples, just the number expected. All ordered pairs $(\infty,(x,i)),((x,i),\infty)$ and $((x,i),(y,i))$ appears in triples of part 1. For ordered pair $P = ((x,i),(y,i+1))$, there is $z \in Z_n$ such that $x+y+z = 0$, then P appears in triple $((x,i),(y,i+1),(z,i-1))$ of part 2. For $P = ((x,i),(y,i-1))$, let $u = f(x), v = f(y), w = -(u+v)$ and $z = f(w)$. Since $f(u) = f^2(x) = x$ and $f(v) = f^2(y) = y$, P appears in triple $((x,i),(y,i-1),(z,i+1))$ of part 3.

Triple $A = ((x,i),(y,i),(z,i)) \in \mathcal{B}$ implies $(x,y,z) \in \mathcal{A}$, so $(f(z), f(y), f(x)) \in \mathcal{A}$, therefore $((f(z),i), (f(y),i), (f(x),i)) \in$ part 1, i.e. $F(A)^{-1} \in \mathcal{B}$. Triple $B = ((x,0), (y,1), (z,2)) \in \mathcal{B}$ implies $x+y+z = 0$, so $((f(z),2), (f(y),1),(f(x),0)) \in$ part 3, i.e. $F(B)^{-1} \in \mathcal{B}$. For the triple $C = ((x,2),(y,1),(z,0)) \in \mathcal{B}$, let $u = f(x), v = f(y)$ and $w = f(z)$, then $f(u) = f^2(x) = x$, $f(v) = f^2(y) = y$ and $f(w) = f^2(z) = z$. Thus,

$C = ((x,2),(y,1),(z,0)) = ((f(u),2),(f(v),1),(f(w),0)) \in B$ implies $u+v+w=0$ and $((w,0),(v,1),(u,2)) = ((f(z),0),(f(y),1),(f(x),2)) \in$ part 2, that is $f(C)^{-1} \in \mathcal{B}$.

Finally, since $f(0)=0$ and $0+0+0=0$, thus $F(0,i)=(0,i), i \in Z_3$ and the $(X,\mathcal{B},f)$ contains a sub-$SCDTS(3) = \{((0,0),(0,1),(0,2)),((0,2),(0,1),(0,0))\}$ by part 2 and part 3. For any element $(x,i) \in Z_n \times Z_3$, $F^2(x,i) = (f^2(x),i) = (x,i)$, thus $p_F((x,i)) \le 2$. The element ∞ is another fixed point. Thus, $(X,\mathcal{B},F)$ is C-type. □

Lemma 7.4.10. If there exists a B-type $SCDTS(n+4)$ then there exists a C-type $SCDTS(3n+4)$, for $n>0$.

Construction. Given a B-type $SCDTS(n+4) = (\{u,v,s,t\} \cup Z_n, \mathcal{A}, f)$, which contains a sub-$SCDTS(4)$ on the subset $\{u,v,s,t\}$, and satisfies $p_f(x) \le 2$ for any $x \in Z_n$, where $x=u,v$ and 0 are fixed points. Construct a transitive triple system $\mathcal{B}$ on set $X = \{u,v,s,t\} \cup (Z_n \times Z_3)$ as follows:

part 1. sub-$SCDTS(4)$ on $\{u,v,s,t\}$, say D.

part 2. $((x,i),(y,i),(z,i))$ with $(x,y,z) \in \mathcal{A} \setminus D$ and $i \in Z_3$. (whenever u,v,s,t appears for x,y,z, omit the second coordinate i.)

part 3. $((x,0),(y,1),(z,2))$ and $((f(z),2),(f(y),1),(f(x),0))$ with $x,y,z \in Z_n$ and $x+y+z=0$.

Define $F(u)=f(u), F(v)=f(v), F(s)=f(s), F(t)=f(t)$ and $F(x,i) = (f(x),i)$ for any $(x,i) \in Z_n \times Z_3$. Then $(X,\mathcal{B},F)$ is an $SCDTS(3n+4)$ expected.

Proof. It is not difficult to show that $(X,\mathcal{B})$ is an $DTS(3n+4)$, the proof is similar to that of Lemma 7.4.9. If B is a triple from $\{u,v,s,t\}$, then $F(B)^{-1} = f(B)^{-1}$ appears in part 1. As a sub-$SCDTS(4)$ of $\mathcal{A}$, D is also a sub-$SCDTS(4)$ of $\mathcal{B}$, therefore $p_F(z) \le 2$ for $z=u,v,s,t$. Other statements of the proof are almost the same to that in Lemma 7.4.9. □

Lemma 7.4.11. If there exists a C-type $SCDTS(n+3)$ then there exists a C-type $SCDTS(3n+3)$.

Construction. Given that a C-type $SCDTS(n+3) = (\{a,b,c\} \cup$

Z_n, $\mathcal{A}, f)$, which contains a sub-$SCDTS(3)$, $D = \{(a,b,c),(c,b,a)\}$, and satisfies $p_f(x) \leq 2$ for any $x \in Z_n$. Let us construct a transitive triple system $\mathcal{B}$ on set $X = \{a,b,c\} \cup (Z_n \times Z_3)$ as follows:

part 1. D.

part 2. $((x,i),(y,i),(z,i))$ with $(x,y,z) \in \mathcal{A} \setminus D$ and $i \in Z_3$. (Whenever a,b,c appears for x,y,z, omit the second coordinate i.)

part 3. $((x,0),(y,1),(z,2))$ and $((f(z),2),(f(y),1),(f(x),0))$ with $x,y,z \in Z_n$ and $x+y+z=0$

Define $F(a) = a, F(b) = b, F(c) = c$ and $F(x,i) = (f(x),i)$ for any $(x,i) \in Z_n \times Z_3$. Then $(X, \mathcal{B}, F)$ is an $SCDTS(3n+3)$.

Proof . Similar to Lemma 7.4.9 and Lemma 7.4.10. □

Lemma 7.4.12. There exists a C-type $SCDTS(v)$ for $v \equiv 4 \pmod 6$ and $v > 4$.

Proof. Let $v = 6k+4$, all possibilities are classified as follows.

Case 1. $k = 3t$, $t \geq 1$, then $v = 18t+4 = 3 \times 6t + 4$.

There exists a B-type $SCDTS$ $(6t+4)$ by Theorem 7.4.3, thus there exists a C-type $SCDTS$ $(3 \times 6t + 4)$ by Lemma 7.4.10.

Case 2. $k = 3t+1$, then $v = 18t+10 = 3 \times (6t+3)+1$. There exists an A-type $SCDTS(6t+3+1)$ by Theorem 7.4.3, thus there exists a C-type $SCDTS(3 \times (6t+3)+1)$ by Lemma 7.4.9.

Case 3. $k = 3t+2$, then $v = 18t+16$.

subcase 1. $t = 2s$, then $v = 36s+16 = 4(9s+4)$. There exists a C-type $SCDTS(9s+4)$ by Theorem 7.4.1 (when s is odd) or Case 1 (when s is even) above, therefore there exists a C-type $SCDTS(4(9s+4))$ by Lemma 7.4.6.

subcase 2. $t = 2s+1$, then $v = 36s+34 = 3(12s+11)+1$. If there exists an A-type $SCDTS(12s+12)$, then there exists a C-type $SCDTS(v)$ by Lemma 7.4.9. we now discuss the existence of A-type $SCDTS(12s+12)$. Firstly, since there exist C-type $SCDTS(18p+4)$ and $SCDTS(18p+10)$ by Case 1 and Case 2, thus, by Lemma 7.4.11, there exist C-type

$$SCDTS(54p+6) \text{ and } SCDTS(54p+24). \qquad (*)$$

Now, we can get the existence of A-type $SCDTS(12s+12)$ (classified by $s = 6m + i$, $i \in Z_6$):

(i) $72m + 12 = 12(6m + 1)$: by Theorem 7.4.1 and Corollary 7.4.7.

(ii) $72m + 24$, let $m = 3p + i$ ($i \in Z_3$):

$216p + 24 = 4(54p + 6)$	by (*) and Lemma 7.4.6;
$216p + 96 = 4(54p + 24)$	by (*) and Lemma 7.4.6;
$216p + 168 = 54(4p + 3) + 6$	by (*).

(iii) $72m + 36 = 4(18m + 9)$: by Theorem 7.4.1 and Lemma 7.4.6.

(iv) $72m + 48 = 12(6m + 4)$: by Theorem 7.4.3 and Corollary 7.4.7.

(v) $72m + 60 = 4(18m + 15)$: by Theorem 7.4.1 and Lemma 7.4.6.

(vi) $72m + 72 = 4(18m + 18)$: by Theorem 7.4.8 and Lemma 7.4.6.

This completes the proof. □

7.4.4 The conclusion

Given that $SCDTS(v) = (X, \mathcal{B}, f)$, for any triple $B \in \mathcal{B}$, define

$$B_0 = B \text{ and } B_i = f(\mathcal{B}_{i-1})^{-1}, \quad i = 1, 2, \ldots$$

If $B_k = B_0$ and $B_s \neq B_0$ for any $s < k$, then denote $k = p_f(B)$. Certainly, when $p_f(B) = k$, $B_0, B_1, \ldots, B_{k-1}$ belong to the system $\mathcal{B}$ and are pairwise distinct. Denote $P(B) = \{B_0, B_1, \ldots, B_{k-1}\}$ if $p_f(B) = k$. Suppose $B_0 = B = (x, y, z)$, then $B_{2i} = (f^{2i}(x), f^{2i}(y), f^{2i}(z))$ and $B_{2i+1} = (f^{2i+1}(z), f^{2i+1}(y), f^{2i+1}(x))$. Thus, it is not difficult to see that $p_f(B)$ is odd only if $p_f(y)$ is odd and both $p_f(x)$ and $p_f(z)$ are even. We can get

fact 1 For $B = (x, y, z) \in \mathcal{B}$, $p_f(B) = l.c.m.\{2, p_f(x), p_f(y), p_f(z)\}$ except when there are two even and one odd among $p_f(x), p_f(y)$ and $p_f(z)$.

For $v = 6$, we have still the following facts:

fact 2 No $DTS(6)$ can contain a sub-$DTS(3)$.

Suppose $DTS(6) = (\{a, b, c, 1, 2, 3\}, \mathcal{B})$, and $(a, b, c), (c, b, a) \in \mathcal{B}$. Then for each $x \in \{a, b, c\}$, $\mathcal{B}$ must contain sub-$DTS(4)$ on the set $\{x, 1, 2, 3\}$ without a triple from the set $\{1, 2, 3\}$. But, it is impossible.

fact 3 If there exists an $SCDTS(6) = (X, \mathcal{B}, f)$, then there is no such $(x, y, z) \in \mathcal{B}$ that x, y and z are all fixed under f.

Lemma 7.4.13. There is no $SCDTS(6)$.

Proof. Suppose there exists an $SCDTS(6) = (X, \mathcal{B}, f)$, where X is a 6-set. The mapping f, as a permutation on the set X, can be denoted by the product of some disjoint cyclic permutations. In what follows, a cyclic permutation with length t will be called *t-cycle*.

(1) It is impossible for f to be an identity mapping by fact 3.

(2) If f contains a 2-cycle, let $f = (a, b)\cdots$. Consider the triple B containing ordered pair (a, b), let x be the third element in B. Then,

$$\begin{array}{rlll} B_0 = B: & (a,b,x) & \text{or } (x,a,b) & \text{or } (a,x,b), \\ B_1 = f(B)^{-1}: & (f(x),a,b) & \text{or } (a,b,f(x)) & \text{or } (a,f(x),b). \end{array}$$

Since each pair appears only in a unique triple, thus $B_0 = B_1$. Therefore, $B = (a, x, b)$ and x is a fixed point, which implies that f must contain a 1-cycle at least.

(i) When $f = (a, b)(x)(u, v, w)$, there are both (a, x, b) and (b, x, a) in $\mathcal{B}$, i.e. a sub-$DTS(3)$, which contradicts the fact 2.

(ii) When $f = (a, b)(1)(2)(3)(4)$, besides the two triples containing both a and b, and the six triples containing either a or b but not both, the system $\mathcal{B}(|\mathcal{B}| = 10)$ must contain two triples from the set $\{1, 2, 3, 4\}$, which will contradicts the fact 3.

(iii) When $f = (a, b)(c, d)(1)(2)$, any triple in $\mathcal{B}$ must contain fixed point 1 or 2 by the above-mentioned. But, all ordered pairs from the set $\{a, b, c, d\}$ (total 4×3=12) will occupy at least 12 triples, it is impossible.

(3) If f contains a 3-cycle, let $f = (a, b, c)\cdots$. Since f can not contain 2-cycle by (2), thus the third element x in the triple B containing ordered pair (a, b) must satisfy $p_f(x) = 1$ or 3. Thus, $p_f(B) = 6$ by fact 1. It is not difficult to see $B = B_0 = B_3^{-1}$, i.e. $\{B, B_3\}$ is a sub-$DTS(3)$, a contradiction with fact 2.

(4) If f contains a 4-cycle, then f must be such form $(a, b, c, d)(s)(t)$ by (2). Consider the triple B containing the ordered pair (s, t). Since the third element x in B must satisfy $p_f(x) = 4$, so $p_f(B) = 4$, by fact

1. Thus $B = B_0, B_1, B_2, B_3 \in \mathcal{B}$. The ordered pair (s,t) appears in both triple B and B_2, it is impossible.

(5) If $f = (a,b,c,d,e)(s)$, then $p_f(B) = 10$ for the triple B containing the ordered pair (a,b), by fact 1. Since $|\mathcal{B}| = 10$, thus B contains element s. Therefore, s appears in all triples of $P(B)$. The pairs containing s are too much!

(6) If $f = (1,2,3,4,5,6)$, consider the triple B, in which the ordered pair (1,4) appears. Since $p_f(B) = 6$ by fact 1, thus $P(B) = \{B_0, B_1, ..., B_5\} \subset \mathcal{B}$. But the pair (1,4) contains in both triples $B_0 = B$ and B_3, it is impossible. □

Theorem 7.4.14. There exists an $SCDTS(v)$ if and only if $v \equiv 0$ or 1 (mod 3) and $v \neq 6$.

Proof . The necessity is obvious, since the spectrum of $DTS(v)$ is $v \equiv 0$ or 1 (mod 3), and there is not $SCDTS(6)$ by Lemma 7.4.13. When $v \equiv 1$ or 3 (mod 6) and $v \equiv 4$ (mod 6), the sufficiency holds by Theorem 7.4.1 and 7.4.3. When $v \equiv 0$ (mod 6), there are three cases to consider: if $v = 18t$ the existence of an $SCMTS(v)$ follows from Theorem 7.4.8; if $v = 18t + 6$, $t \geq 1$, there exists a C-type $SCMTS(6t + 4)$ by Lemma 7.4.12, thus there exists an $SCDTS(18t + 6)$ by Lemma 7.4.11; if $v = 18t + 12 = 3(6t + 4)$, then the conclusion follows from Theorem 7.4.1, Theorem 7.4.3 and Lemma 7.4.2. □

Recently, Tian Zihong obtains the existence spectrum of $SCDTS(v, \lambda)$ without repeated triples for any index λ.

References

[BJL] T. Beth, D. Jungnickel and H. Lenz, Design Theory, Bibliographisches Institut, Zurich, 1985.

[BerFa] J. C. Bermond and V. Faber, Decomposition of the complete directed graph into k-circuits, J. Combin. Theory B, 21 (1976), 146–155.

[Ber 1] J. C. Bermond, An application of the solution of Kirkman's schoolgirl problem: the decomposition of the symmetric oriented complete graph into 3-circuits, Discrete Math., 8 (1974), 301–304.

[Ber 2] J. C. Bermond, Cycles dans les graphes et G-configurations, Thesis, Univ. of Paris XI (Orsay), April 1975.

[Brou 1] A. E. Brouwer, Optimal Packings of $K_4's$ into K_n, J. Combin. Theory A, 26(1979), 278–297.

[Brou 2] A. E. Brouwer, A series of separable designs with application to pairwise orthogonal Latin squares, Europ. J. Combin., 1(1980), 39–41.

[Cay] A. Cayley, On the triadic arrangements of seven and fifteen things, London Edinburgh and Dublin Philos. Mag. and J. Sci., (3), 37 (1850), 50–53 (Collected Mathematical papers I, 481–484).

[Chang 1] Chang Yanxun, On the spectrum of Golf designs, Journal of Hebei Normal College, 2(1992), 8–15.

[Chang 2] Chang Yanxun, On large sets of directed triple systems (to appear).

[ChangKang] Chang Yanxun and Kang Qingde, On the large sets of pairwise disjoint transitive triple system, Chinese Science Bulletin, 36(1991), 1278.

[ChangKangLei] Chang Yanxun, Kang Qingde and Lei Jianguo, The existence of LD-designs and LD^*-designs, Acta Matnematicae Applicatae Sinica, 14(1991), 565–567.

[ChangYangKang] Chang Yanxun, Yang Guihua and Kang Qingde, The spectrum of self-converse MTS, to appear in Ars Combin..

[Chen] Chen Demong, private communication.

[ChenLei] Chen Demong and Lei Jianguo, Some constructions of large sets of disjoint 3-GDDs using prime numbers (preprint).

[ChenLinSt] Chen Demong, C. C. Lindner and D. R. Stinson, Further results on large sets of disjoint group-divisible designs, to appear in Discrete Math..

[ChenSt 1] Chen Demong and D. R. Stinson, Recent results on combinatorial constructions for threshold schemes, Australasian J. of Combinatorics, 1(1990), 29–48.

[ChenSt 2] Chen Demong and D. R. Stinson, On the construction of large sets of disjoint group-divisible designs, Ars Combinatoria (to appear).

[ChenSt 3] Chen Demong and D. R. Stinson, More constructions for large sets of disjoint GDDs (preprint).

[ColCol] C. J. Colbourn and M. J. Colbourn, Disjoint cyclic Mendelsohn triple systems, Ars Combin., 11 (1981), 3–8.

[ColHoRe] C. J. Colbourn, D. G. Hoffman and R. Rees, Group divisible designs with block size three (to appear).

[ColPulRo] C. J. Colbourn, W. R. Pulleyblank and A. Rosa, Hybrid triple systems and cubic feedback sets, Graphs Combin. 5(1989), 15–28.

[ColRo] C. J. Colbourn and A. Rosa, Directed and Mendelsohn triple systems, in Chapter 4 of Contemporary Design Theory, A Wiley Interscience Publication (1992), 97–136.

[Den 1] R. H. F. Denniston, Doubly resolvabilities of some complete 3-designs, manuscripta math., 12(1974), 105–112.

[Den 2] R. H. F. Denniston, Some packings with Steiner triple systems, Discrete Math., 9(1974), 213–227.

[Den 3] R. H. F. Denniston, Sylvester's problem of the 15 schoolgirls, Discrete Math., 9(1974), 229–233.

[Den 4] R. H. F. Denniston, Four doubly resolvable complete 3-designs, Ars Combinatoria, 7(1979), 265–272.

[Den 5] R. H. F. Denniston, Further cases of double resolvabilities, Journal of Combinatonrial Theory *A*, 26(1979), 298–303.

[Do] J. Doyen, Constructions of disjoint Steiner triple systems, Proc. Amer. Soc., 32(1972), 409–416.

[GoTa] S. W. Golomb and H. Taylor, Tuscan squares— a new family of combinatorial designs, Ars Combin., 20-B (1985), 115–132.

[Ha] H. Hanani, On quadruple systems, Canad. J. Math., 12(1960), 145–157.

[Hein] K. Heinrich, Simple direct constructions for hybrid triple designs, Discrete Math. (to appear).

[HsuKeed] D. F. Hsu and A. D. Keedwell, Generalized complete mappings, neofields, sequenceable groups and block designs II, Pacific J. of Math., 117(1985), 291–312.

[HungMen] S. H. Y. Hung and N. S. Mendelsohn, Directed triple systems, J. Combin. Theory A, 14(1973), 310–318.

[Kang 1] Kang Qingde, A recurrence construction of LD-designs, Acta Mathematicae Applicatae Sinica, 10(1987), 481–490.

[Kang 2] Kang Qingde, On large sets of block designs, Ph.D.Thesis, Eindhoven Univ. of Technology, The Netherlands, August 1989.

[Kang 3] Kang Qingde, A generalization of Mendelsohn triple systems, Ars Combinatoria, 29c(1990), 207–215.

[Kang 4] Kang Qingde, The construction of large sets of disjoint Mendelsohn triple systems of order $2^n + 2$, Discrete Math., 90 (1991), 199–205.

[KangChang 1] Kang Qingde and Chang Yanxun, Symmetric Mendelsohn triple systems and large sets of disjoint Mendelsohn triple systems, Combinatorial designs and applications (Lecture Notes in Pure and Applied Mathematics, Vol.126), 69–78, Marcel Dekker Inc., 1990.

[KangChang 2] Kang Qingde and Chang Yanxun, Further results for the large sets of disjoint Mendelsohn triple systems I, Discrete Math., 118 (1993), 263–268.

[KangChang 3] Kang Qingde and Chang Yanxun, Further results for the large sets of disjoint Mendelsohn triple systems II, Chinese Science Bulletin, 37(1992), 90–91.

[KangChang 4] Kang Qingde and Chang Yanxun, A completion of the spectrum for large sets of disjoint transitive triple systems, Journal of Combin. Theory A, 60(1992), 287–294.

[KangChangYang] Kang Qingde, Chang Yanxun and Yang Guihua, The Spectrum of self-converse DTS, to appear in J. of Combin. Designs.

[KangLei 1] Kang Qingde and Lei Jianguo, A completion of the spectrum for large sets of disjoint Mendelsohn triple systems, Bulletin of the ICA, 9(1993), 14–26.

[KangLei 2] Kang Qingde and Lei Jianguo, On large sets of disjoint hybrid triple systems, to appear in JSPI.

[KangLeiChang] Kang Qingde, Lei Jianguo and Chang Yanxun, The spectrum for large sets of disjoint Mendelsohn triple system with any index, to appear in J. of Combin. Designs.

[Kir 1] T. P. Kirkman, On a problem in combinations, Cambridge and Dublin Math. J., 2(1847), 191–204.

[Kir 2] T. P. Kirkman, On the puzzle of the fifteen young ladies, London, Edinburgh and Dublin Philos. Mag. and J. Sci., (4), 23 (1862), 198–204.

[KLR] A. Kotzig, C. C. Lindner and A. Rosa, Latin squares with no subsquares of order two and disjoint Steiner triple systems, Utilitas Math., 7 (1975) 287–294.

[Lin 1] C. C. Lindner, On the number of disjoint Mendelsohn triple systems, J. Combin. Theory A, 30 (1981), 326–330.

[Lin 2] C. C. Lindner, Construction of large sets of pairwise disjoint transitive triple systems II, Discrete Math., 65 (1987), 65–74.

[LinSt 1] C. C. Lindner and A. P. Street, Construction of large sets of pairwise disjoint transitive triple systems, Eur. J. Combin., 4(1983), 335–346.

[LinSt 2] C. C. Lindner and A. P. Street, Ordered triple system and transitive quasigroups, Ars Combin., 17(1984), 297–306.

[LeiKang] Lei Jianguo and Kang Qingde, On large sets of disjoint 3-GDD, Journal of Hebei Normal College, 2(1993), 14–21.

[Lu 1] Lu Jiaxi, On large sets of disjoint Steiner triple systems I, J. Combin. Theory A, 34(1983), 140–146.

[Lu 2] Lu Jiaxi, On large sets of disjoint Steiner triple systems II , J. Combin. Theory A, 34(1983), 147–155.

[Lu 3] Lu Jiaxi, On large sets of disjoint Steiner triple systems III, J. Combin. Theory A, 34(1983), 156–182.

[Lu 4] Lu Jiaxi, On large sets of disjoint Steiner triple systems IV, J. Combin. Theory A, 37(1984), 136–163.

[Lu 5] Lu Jiaxi, On large sets of disjoint Steiner triple systems V, J. Combin. Theory A, 37(1984), 164–188.

[Lu 6] Lu Jiaxi, On large sets of disjoint Steiner triple systems VI, J. Combin. Theory A, 37(1984), 189–192.

[Lu 7] Lu Jiaxi, On large sets of disjoint Steiner triple systems VII, unfinished manuscript.

[Men] N. S. Mendelsohn, A natural generalization of Steiner triple systems, Computers in Number Theory, 323–338, Academic Press, New York, 1971.

[Qui] D. Quiring, A constrction of disjoint Steiner triple systems, J. Combin. Theory A, 27(1979), 407–408.

[Rei] M.Reiss, Uber eine Steinersche combinatorische Aufgabe, welche in 45sten Bande dieses Journals, Seite 181, destellt worden ist, J. reine angew. Math., 56(1859), 326–344.

[Rob] D. F. Robinson, Constructing an annual round-robin tournament played on neutral grounds, Math. Chronicle, 10(1981), 73–82.

[Ro 1] A. Rosa, A theorem on the maximum number of disjoint Steiner triple systems, J. Combin. Theory A, 18(1975), 305–312.

[Ro 2] A. Rosa, Intersection properties of Steiner systems, Ann. Discrete Math., 7(1980), 115–128.

[Sch] S.Schreiber, Covering all triples on n marks by disjoint Steiner systems, J. Combin. Theory A, 15(1973), 347–350.

[SebSki] J. R. Seberry and D. Skilicorn, All directed BIBDs with $k = 3$ exist, J. Combin. Theory A, 29(1980), 244–248.

[So] D. Sotteau, Decompositon of K_n^* into circuits of odd length, Discrete Math., 15(1976), 185–191.

[St] J.Steiner, Combinatorische Aufgabe, J.reine angew. Math., 45(1853), 181–182.

[Sy] J.J.Sylvester, On the fifteen young ladies problem, Proc. London. Math. Soc., 7(1875-76) 235–236.

[Teir 1] Luc Teirlinck, On the maximum number of disjoint Steiner triple systems, Discrete Math., 6(1973), 299–300.

[Teir 2] Luc Teirlinck, On the maximum number of disjoint triple systems, J. Geometry, (6), 1 (1975), 93–96.

[Teir 3] Luc Teirlinck, On large sets of disjoint quadruple systems, Ars Combin., 17 (1984), 173–176.

[Teir 4] Luc Teirlinck, On the large sets of disjoint ordered designs, Ars Combin., 25(1988), 31–37.

[Teir 5] Luc Teirlinck, Loccally trivial t-designs and designs without repeated blocks, Discrete Math., 77(1989), 345–356.

[Teir 6] Luc Teirlinck, A completion of Lu's determination of the spectrum for large sets of disjoint Steiner triple systems, J. Combin. Theory A, 57(1991), 302–305.

[Teir 7] Luc Teirlinck, Large sets of disjoins designs and related structures, in Chapter 12 of Contemporary Theory, A wiley Interscience Publication (1992), 561–592.

[Teir 8] Luc Teirlinck, Large sets with holes (preprint).

[Teir 9] Luc Teirlinck, On the use of pairwise balanced designs and closure spaces in the construction of structures of degree at least 3, Le Matematiche, Vol XLV (1990), Fasc. J, 197–218.

[TeirLin] Luc Teirlinck and C. C. Lindner, The construction of large sets of idempotent quasigroups, Eur. J. Combin., 9 (1988), 83–89.

[Til] T. W. Tillson, A Hamilton decomposition of K^*_{2m}, $2m \geq 8$, J. Combin. Theory B, 29(1980), 68–74.

[Wal] W. D. Wallis, The problem of the hopitable golfer, Ars Combin., 15(1983), 149–152.

[Wil 1] R. M. Wilson, Some partittions of all triples into Steiner triple systems, in: Hypergraph Seminar (Ohio State University 1972), Lecture Notes in Math. 411, Springer, Berlin 1974, 267–277.

[Wil 2] R. M. Wilson, Construction and uses of pairwise balanced designs, Mathematical centre Tracts, 55(1974), 18–41.

[Wil 3] R. M. Wilson, Decomposition of complete graphs into Subgraphs isomorphic to a given graph, in Proc. 5th British Combinatorial Conference, Aberdeen, July 1975, Congressus Numerantium 15, Utilitas Math., 647–659.

[Wo] W. S. B. Woolhouse, Prize question 1733, Lady's and Gentleman's Diary (1844).

[Wu] Wu Lisheng, On large sets of disjoint Mendelsohn triple systems, J. Suzhou Univ., (2), 3 (1987), 1–4.

[YangKang] Yang Guihua and Kang Qingde, On large sets of $MD(v,k,2)$ (preprint).

Appendix

A Biographical Sketch of Lu Jiaxi

Lu Jiaxi was born on June 10, 1935 in Shanghai and died on October 31, 1983 in Baotou, the Inner Mongolia Autonomous Region. Lu began his schooling at the age of five. He went to Zhengde Elementary School on Nanxun Road in Shanghai and Shengyang Secondary School and then transferred to Mellon Secondary School (today's name: Jiguang Secondary School). Then he couldn't afford to continue his schooling after graduation because of impoverishment and was apprenticed in a hardware store. After Shanghai was liberated in May, 1949, Lu was admitted to a statistical training class by Electrical Appliance Industry Administration in Northeast China. After a short period of training, he was assigned in May, 1952 to work as a statistician at the production section in Harbin Electrical Machinery Plant in Heilongjiang Province in Northeast China. He got the chance of receiving higher education and became a student in September, 1957 majoring in physics at Northeast Teachers University. He graduated in 1961 and was assigned to work as an assistant at Baotou Steel-Iron Institute in Inner Mongolia in North China. He was transferred to work as a teacher in the department of education in Baotou when the institute was dismissed afterwards as a result of readjustment of universities and colleges. Lu first worked at the teaching and research section of Baotou Education Bureau and afterwards at No.8 Secondary School, No.5 Secondary School, No.24 Secondary School and last, taught physics at No.9 Secondary School till he died a tragic death in October of 1983 at the age of 48.

It seems that he had nothing to do with mathematics all his life, but he fully deserved a specialist in combinatorial mathematics. He was really the only amateur mathematician with a brilliant contribution among mathematicians in modern China.

Lu Jiaxi was born from a poor family. His father was a pedlar with meagre income. His mother had to help support their family by doing washing and sewing for better-off families. Lu, the only child of his parents, treasured the opportunity of going to school given by his hard-working and thrifty parents. He worked very hard and did exceedingly well in his studies at school. Lu discontinued his studies at school after his father's death and still diligently studied on his own. He resolved to win honour for his country by scaling new heights in science in the future. While in Harbin Etectrical Machinery Plant, he self-studied subjects offered in senior high school and Russian in his spare time. From then on he began to do cursory reading on astronormy, geography, literature and art, philosophy, ethics and other subjects.

In this period, something happened which influenced his lifetime. Once he got a copy of *An Introduction to Interesting Mathematics Methods*, a popular edition of mathematics written by China's senior mathematician Sun Zeying. When he read it he was fascinated by the "Kirkman's schoolgirls problem" and "problems on Steiner systems" introduced in the book, and he was keen on a try to solve the two problems. From 1957, the year before his college life, he channeled it into his life trying to solve the two problems in combinatorial design. Mathematics was not his field. At college he majored in physics and taught it at high schools. Although mathematics was one of the basic courses he studied at college, it was far from being enough for him to be engaged in mathematical research. Moreover, combinatorial mathematics had not been systematically studied in China in the late 1950s. At that time he could get neither reference books in Chinese nor help from specialists. It was sheerly by his curiosity he forced his way in the field of the problems of combinatorial design in such poor circumstances.

As the saying goes, everything is hard in the beginning. Lu Jiaxi was an layman for mathematics. It was hard to imagine his first step in his research. For some smaller orders, "Kirkman's schoolgirls problem" might be regarded as a sort of digital game, and its solution be pursued with the method of piecing together. Those (including Lu himself) who first touched the question of combinatorial design tended to begin to research into it with interest. As the order becomes bigger, one cannot win a success on the design problem through good luck. It seemed that to get the correct configuration among many (even astronomical) possible arrangements was no easier than to fish a needle out of the ocean. Even a modern high-speed computer is far from doing that. The key here is to master the methods and tools of modern design theory.

Lu Jiaxi realized that he must enrich himself. Once he talked about his studying and research methods in a speech at a conference in Dalian in 1983 before he died. He remarked, "Thanks to the small book by Mr. Sun Zeying, which was an interesting guide for me, and some other professional books which catered for readers of various levels, I have been approaching forward position of combinatorial design from a layman for mathematics. As a beginner, I didn't have much time to study those thick professional works. I studied practically with questions in my mind. And I am grateful for the work of other mathematicians at home and abroad. Their articles and works have been the cornerstone of my achievements and as a matter of fact I have borrowed some good methods from them...."

After being admitted into college, industrious Lu Jiaxi did not appear to be reluctant to borrow from the library some most important literature which dealt directly with the "Kirkman's schoolgirls problem". He began to study those mathematical concepts, terms and methods which he hadn't understood and study the methods of design theory from the professional works. He attached importance on practising while studying and then meditated on the problem. Later he became aware of the benefits from practice and that made him full of confidence when he discovered a new problem, he set himself to further studies. To solve the design problems, he studied succesively Modern Algebra, Elementary Number Theory, $(0, 1)$-matrix Theory, Finite Geometry, Difference set, Orthogonal Latin squares and so on.

After class and finishing his assignments, Lu threw himself into his small mathematical field — reading up the literature in the library and consulting senior teachers and specialists. He made full use of his spare time. When electricity was cut off at the bed time, he got out of the room and performed mathematical calculations by the dim light of a corridor light. When his classmates went out on excursion on weekends, he hid himself into a guiet place to do his meditating. Nothing in the world is difficult for one who sets his mind on it. When he graduated four years later, he was not only granted a diploma on physics owing to his excellent marks but also finished his first article on mathematics which was his first step on the way to the solution of the Kirkman's schoolgirls problem.

Lu was very busy with his job when he became a teacher of middle school after his graduation. Sometimes he took three senior classes, which involved five lesson plans and fifteen periods a week. Even he had to take on teaching tasks and experiment of two junior classes and another two senior classes of a high school. But he didn't stop his meditation on the two mathematical problems. He felt that he had to race against time and often worked overtime. He developed the habit of working late into the night and sometimes worked through the night. He still worked on Sundays and during holidays and even over the Spring Festival which the Chinese people enjoy very much and have a three-day off work in February of the year. For Lu, two *mantou* (steamed bread) could serve as a meal. He was sometimes too busy with his job and research to go to the barber's. He went so far as to put off his marriage till he was 37.

Albert Einstein once commented on some people who was engaged in scholarship: "On no account must I stand such scientists who get a board and bore a hole in it at the thinnest point." Lu Jiaxi was not at all such a man as those commented on by Einstein. What he did was not at all to bore a hole in a board, but to "gnaw a backbone or a steel plate" instead. He "gnawed" the mathematical problems with the tough spirit of "water-constantly-dripping-wears-holes- in-stone" (the spirit of "Little strokes fell great oaks"). He dedicated his life to solving the problems.

The people have always expressed admiration for the celebrated Curies. Their laboratory in which they worked and discovered radium was described as follows: "It is in such a bad condition that it looks like a stable or a small vegetable cellar. It is too shabby to be a laboratory." Then how many people can match China's intellectuals both of assiduousness and of endurance? Lu Jiaxi and his wife lived in a small room of four metres by three metres with their two children. It was a bedroom, a kitchen and a study all in one with some obsolete furniture against walls.

Lu had his daughter who was a pupil use the only desk and he himself had to make do by bending over their shabby heatable brick bed patched with old newspapers as he worked. When he considered some combinatorial configurations, he made do with some matches! When he wrote his manuscripts, he used bits of paper of different size because he couldn't afford to buy it. Imagine he earned only 50 or 60 yuan (about $ 10) a month at that time.

The city of Baotou in Inner Mongolia where he lived lies in the border area. It was difficult for Lu to get touch with the latest development in the outside world. He had to keep up a correspondence with some of his acquaintances and friends working in the libraries of colleges and universities. And sometimes he had to go up to Beijing for information and materials at his own expense. Every time he was in Beijing he worked in Beijing Library during the day and went to find a bench in the waiting-room of the railway station to pass the night. He ate the food he brought from home because he couldn't afford to dine in a restaurant. On the other hand he spared no money to buy books on mathematics and in English from bookstores.

In his spare time Lu liked listening to Peking Opera, so he had some records of the opera . But in order to improve both his writing ability and reading comprehension in English, he parted reluctantly with the opera records he treasured and exchanged some English records with them. During the period of time in which he wrote his articles for the American journal *Journal of Combinatorial Theory*, he completed his tasks he regulated for himself in a day and typewrote six extra pages before going to bed. He dedicated all he had — his time, his energy and his money to the subject. When he died, he left his wife and daughter nothing but some piles of manuscripts, a shabby room with nothing valuable in and a debt of several hundred yuan. In their memories, Lu's daughters wrote, "Father, you left us in such a hurry. Just a few years ago you suggested that we take a photograph of the whole family in a photo studio, but you could not afford the time. Now that you are gone to another world we have to keep you and such a photo in our heart."

Lu's wife once talked about him, "When he was a small boy, misfortune befell him and he suffered from a miserable life as a result of his parents's early death. It was our motherland and people that nurtured and educated him. Though he was a recicent man, he often told us that he could not have been through college without the people's scholarship and could not have made the achievements without the support by colleagues. He loved mathematics. He loved his career. He was seemingly crazy about those strange figures and numerals. But the reason why he worked on the mathematical problem day and night was that he thought it more important to make achievements to render service to repay the kindness of his country and people." These remarks really reveal the reticent scholar's inner world. Lu kept his word by hard work and inflexible pursuit — " Work for the mankind sufficiently "

The most pleasant thing for a devoted pursuer of knowledge is no greater than that of the public recognition of his creative work. Nevertheless, Lu Jiaxi had never almost so experience on his way to the suceess. All of his scientific manuscripts written from 1961 to 1983 when he died are as follows:

1. 30/12/1961 Construct orthogonal Latin squares using block designs;
2. 30/12/1961 A method to construct Kirkman and Steiner systems;
3. 14/3/1965 Constructing methods for $BIBD$ and $RBIBD$;
4. 7/8/1965 Acombinatorial method to construct orthogonal Latin squares;
5. 7/8/1965 On $B[k,1;v]$ with prime power k;
6. 22/1/1966 On $B[5,\lambda;v]$;
7. 29/1/1966 A $B[5,1;141]$;
8. 6/5/1978 On Kirkman problem I;
9. 6/5/1978 Combinatorial method to construct Kirkman systems;
10. 22/5/1978 On Kirkman problem II;
11. 2/7/1978 A new view for Kirkman problem;
12. 6/3/1979 On Kirkman quadruple systems;
13. 7/3/1979 A family of distinct Steiner triple systems;

14. 7/5/1979 Disjoint Steiner systems and Bays's conjecture;

15. 20/7/1979 The existence of resolvable balanced incomplete block designs;

16. 18/9/1981 On large sets of disjoint Steiner triple systems I, II;

17. 7/4/1982 On large sets of disjoint Steiner triple systems III;

18. 17/6/1982 On large sets of disjoint Steiner triple systems IV;

19. 4/3/1983 On large sets of disjoint Steiner triple systems V, VI;

20. 5–10/1983 On large sets of disjoint Steiner triple systems VII;

The dates given above were those when he contributed them to magazines. But they were all turned down or were missing for ever — like a stone dropped into the sea except the 15th–19th papers being published in the two famous magazines at home and abroad before and after his death. This is to say, his scientific discoveries had not been recognized by the public until 1983.

In fact, there were some things that accounted for the occurrence: some papers were missing due to the poor social environment (e.g. During the social upheaval of ten years from 1966 to 1976 most of the academic magazines stopped publication.); some of them as such were of great value. but they were overridden by foreign researchers' articles because of poor information and delay; some valuable articles were not appreciated by those editors who hadn't recognised the value of them or had misjudged them. The most unforgiving is the hasty, carless dealing with the the 2nd and 3rd papers. The mistake by the editors unfortunately caused $RB[3,1;v]$ and $RB[4,1;v]$ to be published some years later. Whatever the reasons had been could be a hard unfair blow on Lu. He didn't fail to confront all the frustration and unluckiness. He was not frustrated and didn't give himself up as hopeless. He continued working diligently.

On the other hand Lu suffered some other sorts of unfairness. During the ten years of the extremeleftist trend of thought (1966–1976), Lu had been sneered by somebody. He was referred to as "a fool" or "a lunatic", and they even made unwarranted charges against Lu that he was after fame and wealth and didn't attend to his proper duties. For some time he was just put labels on that he did not care about politics and as a result, he was sent to a cadre school to be criticized and to be reformed through labour. The shortsighted, lame Lu was burden with all this unrecognized achievements in scientific research, poverty in life and being politically depressed. In spite of all this, Lu held out surprisingly and met difficulties indomitably.

"He who knows how to laugh laughs last", as it did in the case of Lu Jiaxi. In July 1983 Lu was invited to give a report on his achievements in scientific research at the first national combinatorial mathematics conference held in Dalian, Liaoning. He was warmly welcomed and won the acclaim of the people present at the conference. The American *Journal of Combinatorial Theory*, one of the authoritative magazines in the world, published six of his papers [Lu 1]—[Lu 6] respectively in the two issues of 1983 and 1984. *Acta Mathematica Sinica*, one of the authoritative magazines in the mathematical world in China, published in full his important paper on the Kirkman problems in the last issue of 1984. South China Teacher University was even ready to transfer him to a teaching post at it in order to make further researches. Toronto University intended to invite him to Canada for co-operative researches. The chief editor of the American *Mathematical Reviews* invited him to be the reviewer of the magazine. The 4th National Conference of the China Mathematics Society held in Wuhan in October, 1983 even made an exception by asking him to give a report as the only representative from a middle school.

At this time death was befalling him unknown to god or ghost. When he got back home on the evening of October 30, 1983, he couldn't help telling his wife about his excitement and his further researching plan. His wife recalled after his death that on the evening of the particulter day she first saw him laugh. He was too excited. He did laugh a hearty laughter. But that was the last laughter of his. Early in the next morning he passed away quietly, unexpectedly and with a smile on his face.

After his death, letters and telegrams of condolence, flooded in. Some mathematicians and scholars, some universities and institutions at home and abroad, expressed condolences on the unexpected death of Lu Jiaxi. China's *Mathematical Studies and Comments* on behalf of the mathematical community in China carried a special article upon Lu's death. All China's mediums reported Lu's deeds and achievements in matheratics. Lu was granted the title "teacher of special classification" by the governments of Baotou City and the Inner Mongolia Autonomous Region.

What Lu Jiaxi contributed to Combinatorics will remain for ever in the history of mathematics.